Climate Change, Shifting Cultivation and Livelihood Vulnerabilities in India

Niranjan Roy · Avijit Debnath · Sunil Nautiyal

Climate Change, Shifting Cultivation and Livelihood Vulnerabilities in India

An Analytical Study

Springer

Niranjan Roy
Department of Economics
Assam University
Silchar, Assam, India

Avijit Debnath
Department of Economics
Assam University
Silchar, Assam, India

Sunil Nautiyal
Centre for Ecological Economics
and Natural Resources (CEENR)
Institute for Social and Economic Change
(ISEC)
Bengaluru, Karnataka, India

ISBN 978-3-031-54926-7 ISBN 978-3-031-54927-4 (eBook)
https://doi.org/10.1007/978-3-031-54927-4

This Springer imprint is published by the registered company Springer Nature Switzerland AG
The registered company address is: Gewerbestrasse 11, 6330 Cham, Switzerland

Paper in this product is recyclable.

Dedicated to the "Jhummia's" of Khonoma village, Nagaland, India for their untiring efforts in making shifting agriculture a sustainable activity.

Foreword

Since the emergence of the Inter-Governmental Panel on Climate Change (IPCC), jointly established by the WMO and UNEP in 1988, the scientific community has been trying to gather information and develop climate policies across the globe. This has ushered in a new beginning in international climate change mitigation treaties and negotiations. The World Commission on Environment and Development report on 'Our Common Future' in 1992 shared a broad spectrum of concerns on global climate change, its implications and mitigation strategies. The comprehensive action strategy 'Agenda 21' focused on creating agreements and conventions on critical issues hovering over climate change, desertification and deforestation. There is a north south divide in specialising conservation efforts to mitigate climate change effects globally. Anthropogenic or human-induced climate change results from human activities leading to deforestation, land use changes, burning fossil fuels, waste management, etc. Since the majority of the workforce depends on agriculture and allied activities in low- and middle-income countries, there is an over-exploitation of natural resources in terms of land use change and deforestation.

Moreover, most indigenous tribal communities depend directly on forest resources, causing changes in the forest ecosystem. Though the indigenous tribal institutions in such countries as India have evolved since antiquity and adapted to man-nature relations through an ideological framework, the higher population growth has led to disastrous consequences on environmental sustainability, as indicated by changes in the forest ecosystem through deforestation and desertification. Shifting agriculture or 'jhumming' is one such practice followed by indigenous tribal communities in India for centuries. This age-old practice has come under scrutiny as it causes large-scale destruction of forest resources in India and livelihood vulnerabilities.

The book explores how climate change affects households that rely on shifting cultivation and how to assess their vulnerability. This book examines indigenous communities practice of shifting cultivation in the micro-level studies conducted in 52 villages and 1469 households covering 7067 populations in seven Northeastern states of India. Most of the studied population belongs to the disadvantaged section of society, living within forest ecosystems and are dependent directly on natural resources for their livelihood.

This book has ten chapters covering different topics related to climate change, such as its patterns, impact on households and agriculture, forest management and the role of indigenous knowledge in mitigating it.

The book has immense potential for policymakers, researchers, climate activists, institutional and non-institutional organisations and people interested in environmental science and social sciences. The book is highly relevant for them as an authentic source on shifting cultivation, climate change and livelihood vulnerability, especially in the Northeastern states of India. This book will inspire future research agendas on climate change and the role of indigenous knowledge systems in their mitigation strategies.

November 2023

Professor Koustuv Dalal, Ph.D.
Editor-in-Chief: Health Economics Review, Springer-Nature,
Senior Editorial Board Member: BMC Health Services Research,
Distinguished Visiting Professor: University of Johannesburg, South Africa, Al-Farabi Kazakh National University, Almaty and many others,
Senior Advisor (Former): WHO CC CSP, Director (Former): Centre for Injury Prevention and Safety Promotion, Support Centre: WHO CC CSP
Senior Health Economist, Faculty Board Member (Human Sciences)
Mid Sweden University
Sundsvall, Sweden

Preface

Shifting cultivation is the modest method available to the tribal population living in the mountainous region. There is no other alternative way of cultivating the land in the mountain areas. It is the natural mode of life for the indigeneous societies in the hilly areas. This system of cultivation is mainly practised in hilly areas of the Northeastern region and other states in India. The indigenous tribal people residing in the hills are arranged socially according to the principles of collective ownership, responsibility and participation. The major advantage of shifting cultivation to the hill people is that it provides a straightforward method of preparing land for cultivation. Weeds and jungles can be quickly cleared by burning process and yields can be obtained within a short period. The destruction of forests, however, is the most significant negative impact of shifting cultivation.

The present study is an excellent insight into the problems of shifting agriculture in India carried out at the micro-level. The field study is conducted in the Eastern Himalayan region of India covering the eight states of the North Eastern Region: Arunachal Pradesh, Assam, Manipur, Meghalaya, Mizoram, Nagaland, Sikkim and Tripura. The field study was carried out in 52 villages, and 1469 households covering 7067 populations in seven states of the region. In addition, many Focused Group Discussions (FGD) were conducted in the selected locations in the region.

Though the percentage of forested area in the Northeastern region is much higher than the country as a whole, due to the rapid growth of the tribal population, forest land has declined significantly. The study addresses a variety of climate change-related topics, including patterns, effects on households and agriculture, forest management and the contribution of indigenous knowledge to its mitigation. This study is associated with different sectors like shifting culture, forestry sector, rural development, etc., and integrated with large respondents and stakeholders through both direct and focus group discussions. The share of the Northeastern region in India's food production has become very insignificant over the years. The age-old practice of shifting agriculture in traditional tribal societies could not transform into modern agriculture. In India's Northeastern region, produce that is dependent on CPR is vital to people's livelihoods.

The study has estimated a livelihood vulnerability index for each of the seven states in Northeastern India using seven indicators: socio-demographic profile, livelihood strategies, social networks, health, food, water, natural disasters and climate variability. Besides the long-term fluctuations of weather variables mainly rainfall and temperature have been captured by using econometric models.

There is a cause-and-effect relationship between climate change and shifting agriculture, as the Northeastern region has long shown signs of extensive forest degradation. The study has found that the indigenous people are aware of the negative effects of shifting cultivation in the light of the possible impact on climate change in many areas of the region. The 'Jhummias' of Khonoma village in Nagaland, India, and their constant efforts to switch over to the settled cultivation by using Aldar tree methods is a glaring example of making a conscious decision to mark shifting agriculture an environment-friendly practice. The study also attempts to address the impact of governance of communal resources based on FGD with indigenous communities. Specific care has been given to the role of community institutions in the sustainable management of forest resources and CPR in the Northeastern region. The role of the Government in its incredible journey towards declaring fully organic farming in Sikkim is another example of institutional effort in the governance of natural resources.

Silchar, India — Niranjan Roy
Silchar, India — Avijit Debnath
Bengaluru, India — Sunil Nautiyal

Acknowledgements

The authors acknowledge deep sense of gratitude to the Indian Council of Social Science Research (ICSSR), New Delhi, Government of India for financial support to conduct the study. The study was based on both the primary and secondary data. The field survey was undertaken in the seven states in the Northeastern region of India. The authors received generous support from the respondents mainly belonging to the indigeneous tribal communities of the Northeastern states. The concerned government officials in the different states extended their helping hand during the field survey. The authors convey special thanks to the Research staffs and Field Investigators from the different states who were involved in collecting primary data. The author expresses deep sense of gratitude to the Assam University authority for providing infrastructural facilities to carry out the study. Also special thanks to the faculty members of the department of Economics for their support and encouragement.

Map 1: India and Its Northeast Region

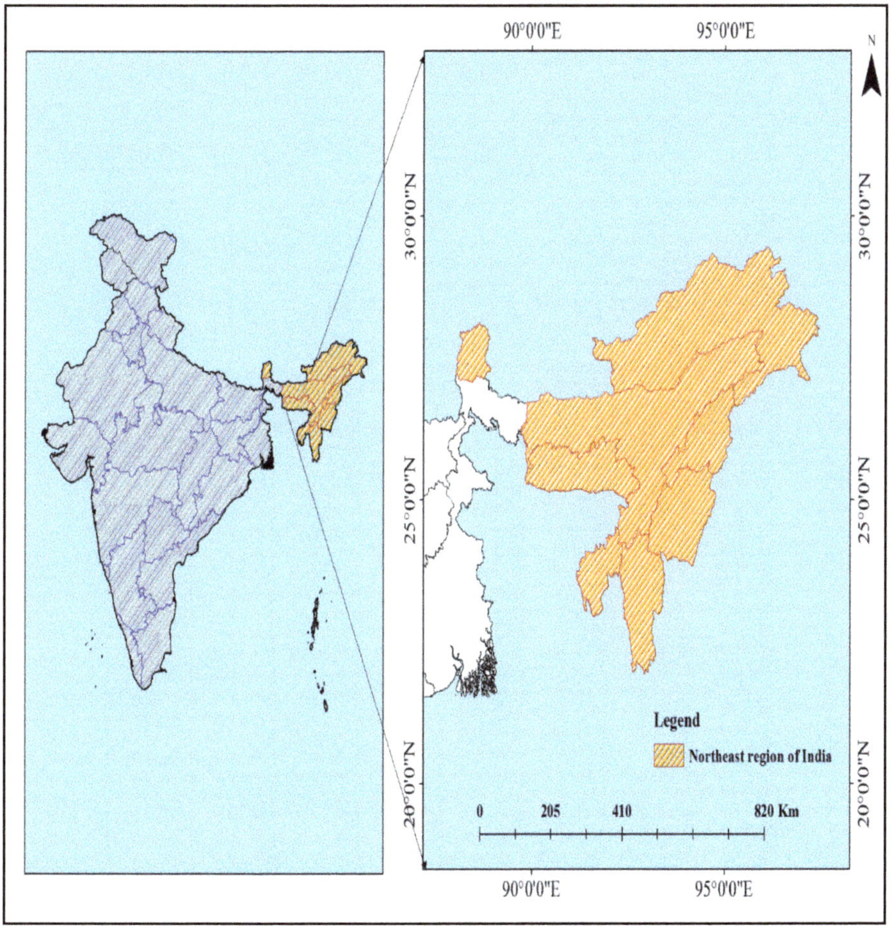

Contents

Abbreviations

CPR	Common Property Resources
FGD	Focus Group Discussion
IPCC	Inter-Governmental Panel on Climate Change
LULC	Land Use and Land Cover
LVI	Livelihood Vulnerability Index
NEC	North Eastern Council
NER	North Eastern Region

Chapter 1
Introduction

The indigenous people who live in the hilly area have access to the most basic technique known as shifting agriculture. There is no other method for farming the soil in the hilly regions. For the tribal people, shifting farming is a natural way of life. Man–nature relationship of a community is reflected in its mode of resource use across the ideological axis. At a given level of ideology, man's material requirements and his perception of nature determines his interaction with nature and characterises an ideal type of the model of resource use with spiritual belief. At this level, ideology acts as a determinant variable of the mode of resource use and is external to the system. The ideological base of a community governs production relations, factors involved in the production and use of natural resources (Gadgil & Guha, 1992). Use of land resources for shifting cultivation by the different indigeneous societies is, however, not indiscriminate, rather it follows certain norms even at an ideological level.

The arrangement of shifting agriculture is mainly practised in hilly areas of North-eastern region and in other few states in India. This type of agricultural practice involves a number of operations. First, forest bushes are cut up to the stump level in the months of December–January, and small trees and bamboos are felled. Short tree stumps and large tree boles are left intact. The stumps are allowed to dry and fire is used to burn. After the burning is completed, the land is cleared, and seeds of different crops as required are planted before the onset of the rains. After every crop season or two, the area is left for several years, and the cultivator moves to another area to slash and burn and cultivate. The jhum cycle used to last as long as 25–30 years. But it has been shortened considerably to 4–5 years in current period due to rapid increases in the tribal population. Long cycles were possible earlier because population pressure was not heavy, and land availability did not limit the process. Shifting cultivation since its inception is identified with the rotation of fields rather than rotation of crops, absence of draught animals and manuring, use of human labour only, a stick, and short period of occupancy alternating with long fallow periods. However, alternating farming is a way of life of the societies practising it. The social organisation of the tribes' active in the mountains is often built around the concepts of

N. Roy et al., *Climate Change, Shifting Cultivation and Livelihood Vulnerabilities in India*, https://doi.org/10.1007/978-3-031-54927-4_1

community ownership, participation and responsibility. To protect the rights of the indigenous tribal population of Assam, Meghalaya, Tripura, and Mizoram, the Sixth Schedule of the Constitution of India given the provision for the administration of these areas. The Articles 244(2) and 275(1) offer this particular provision, which recognises communal ownership or common property resources in the Northeastern region of India.

Agriculture has escaped the mandatory environmental controls applied to other economic sectors, such as application of a polluter-pays principle, pollution taxes and so on. Today, such measures are increasingly mooted as a means to control modern agriculture. Still, widespread changes to landscapes to ecosystem and natural resources have occurred already and ultimately lead to vulnerabilities in livelihood pattern. Initially, the scale and implications of change brought out by agricultural modernisation took environmentalists and policymakers by surprise. In recent decades the environmental impact of intensive agriculture is a significant issue in all countries. The agrarian change and its effects on the environment is a growing concern for all the developed or underdeveloped countries. These concerns have led to new initiatives and policy commitments towards 'sustainable development' which shifts the emphasis from inputs to resource management.

The fact that the hills people can prepare land for cultivation in an easy-to-understand manner is a major benefit of shifting farming. Slash and burn methods can be used to swiftly eradicate weeds and jungles, producing a harvest in a short amount of time. But the most significant negative impact of shifting agriculture is that it causes severe soil erosion from rainfall since it destroys the forest in the hilly areas. The loss of 'topsoil' in hilly regions as a result of shifting farming severely reduces soil fertility, which poses a serious issue for sustaining livelihoods for the indigeneous societies.

In this context the following conceptual framework are very pertinent in understanding the concerns of the climate change and its dynamics related to livelihood vulnerability:

Climate Change

According to WMO climate change refers to long term variations in the weather variables. Human activity is said to be the cause of the current global warming mainly industrial production and land use changes due to agricultural activities. Most developing countries' social and economic systems are projected to suffer significant harm from the effects of climate change (McCarthy et al., 2001). In general, human actions that alter the composition of the atmosphere and seriously harm social, economic, and ecological systems are thought to be the cause of climate change (Brown & Crawford, 2008). For instance, clearing or burning forests increases emissions, which are currently estimated to be between 1.1 and 1.7 billion tonnes per year and mostly harm those with few alternatives for adaptation (Adger et al., 2003). Furthermore, Reid and Huq (2005) define climate change as a substantial and long-lasting shift in the statistical distribution of weather patterns over time. Nevertheless, the definition of climate change in the present study is limited to long-term variations in weather brought on by human activity.

Adaptation

According to Reid and Huq (2005) and Mansourian et al. (2009), adaptation describes both the strategies for adjustments intended to lessen or recover from risks associated with changing climatic and environmental conditions as well as the natural reactions that arise in response to imposed conditions. Moreover, adaptation refers to modes of group behaviour that seek to bring about changes in the ways that society and nature interact.

But according to Lodha (1992), adaptation is a modification of structure, function, or form that results in improved acclimatisation to environmental factors, both cultural and physical. In this study, "adaptation" refers to the coordination of efforts to assist vulnerable rural communities in adjusting to the effects of changing climate conditions. This includes helping them to become aware of and knowledgeable about risk, analyse it, make plans for change, and adopt creative, resilient livelihood systems by identifying and focusing on approaches that take into account their needs and priorities as well as the assets that they have access to.

The Forest

A forest is a complex ecosystem made up of both living things such as trees and birds, and non-living things such as fallen logs, water and rock that dynamic form unit whose composition changes overtime as a result of interacting and living together (Larson et al., 2011). However, according to Bettinger et al. (2009), all living things in the forest are part of a vast, complex and interacting web of life and each living thing has its role in the health and growth of a forest. Meanwhile, Lodha (1993), refers to the forest as a tract of woodland or a type of biome in which the dominant plants are trees. The forest in this study refers to a bunch of trees including all living and non-living things that are closely linked with the meteorological and hydrological ecosystem functioning in an area, i.e. support as game habitat, fire behaviour and tree restoration progressions in the place.

Rural Livelihoods

Chambers and Conway (1992) defined livelihood as the activities, resources, and capacities necessary for an individual's means of subsistence. It also refers to the requirements necessary to maintain people's everyday lives, including the progression of human events and activities. Moreover, human occurrences and activities (processes) that sustain human habitation are referred to as rural livelihoods (Hussein, 2002). In this study, "rural livelihoods" refers to livelihoods associated with forests, namely those related to disaster adaptation, i.e., livelihood diversity and security, as well as creating an environment that makes it possible for people to live in areas near forests on a daily basis. Thus, the term "rural livelihoods" often refers to the set of behaviours necessary for people to live in rural areas.

Local Community

A local community is made up of people who live and work in close proximity to one another. They may exchange resources, knowledge, and support among themselves.

It is a collection of individuals engaging in a shared setting that has intent, beliefs, resources, preferences, wants, hazards, and a number of other circumstances that may be typical in the region and influence the individuals' identities and levels of cohesiveness (Chambers & Conway, 1992). Moreover, a local community refers also to a group or groups of people (individuals, families or households) who share a physical space and sometimes have common interests.

Vulnerability

The degree to which a system is vulnerable to and incapable of adjusting to the negative consequences of climate change, such as climate extremes and variability, is known as its vulnerability. It depends on a system's sensitivity, ability to adapt, and the type, amount, and rate of climatic change and variation to which it is exposed.

1.1 The Northeast India and Dependency on Shifting Agriculture

The Eastern Himalayan region covers the eight states of the North Eastern Region (NER): Arunachal Pradesh, Assam, Manipur, Meghalaya, Mizoram, Nagaland, Sikkim and Tripura. Altogether, the NER covers 7.76% of India's geographical area and is home to two hundred linguistically and culturally diverse inhabitant tribes. The NER is considered as one of the most resourceful regions in India in terms of land and forest (Agarwal, 1996).

The composition, geography, culture and economic structure in the Northeastern region is concerned, there is enormous diversity among the states. A large share of the population of the NER is indigeneous tribal communities. The total population of the region as per 2011 census was 45.8 million. The NER is primarily rural-based where the majority of the population lives in the villages. Majority of the indigeneous societies reside in the hill regions of Mizoram, Nagaland, Meghalaya and Arunachal Pradesh, and form a vast majority of the population of these states. In the state of Arunachal Pradesh, Meghalaya, Mizoram and Nagaland it varies between 85 and 95%. The total literacy rate in NER is higher than the country's average. Mizoram registers the second-highest literacy rate in India. Cultivators and agricultural labourers in the working population consisted of more than 50% of the people as per the 2011 population census.

Hills constitute more than 70 per cent of the total geographical area in NER, where shifting agriculture is the principal method of cultivation (Ramakrishnan, 1996). The ownership pattern of the forest is different in different states of the region. In the hilly areas, the ownership typically rests with the local communities, mainly categorised as Common Property Resources (CPR) (Gadgil & Guha, 1992). Due to the rapid growth of the tribal population, forest land has declined significantly. Therefore, the 'Jhum cycle' is reduced to only 4–5 years. The cultivation is usually uneconomical during the third year and beyond. Also, a considerable number of important tree species, wildlife, and wild plants signifying vast diversity of gene pool are lost due to shifting cultivation (Singh et al., 2003). Changes in the rainfall, temperature and humidity are also taking place. This has consequences for crop yields in NER.

Over time, there has not been much of a progress in the proportion of NER in India's agricultural production. It is pertinent to mention that in India, the technological transformation in the agricultural sector started in the mid-1960s and resulted in 'Green Revolution' in different parts of the country. This transformation of the farming sector has ensured food security in the entire country. The Northeastern states could not take part and adapt the technological transformation process to date. Unfortunately, the whole area still suffers from food insecurities to a great extent. The age-old practice of agriculture in traditional tribal societies has been the natural way of life since antiquity. Ensuring food security is a big concern for the indigenous household.

The Northeastern region India is mostly hilly. It is surrounded by the Himalayan range and landscapes plains on both banks of the Brahmaputra river. This region is home to a wide variety of plants and animals. 7.8% of India's total land area is made up of the region. The northern border of the area is protected by hill ranges that are a part of the Himalayas. The terrain consists of lowlands marginally above sea level and mountains above the snow line. Nearly 90% of the region's borders define India's international borders, making it strategically significant for the nation. The geography of each state in the area is different. Most of Arunachal Pradesh, Mizoram, Nagaland, Meghalaya, almost half of Tripura, one-fifth of Assam, and nine-tenths of Manipur are covered in mountains and hills.

The region's plains are primarily composed of three distinct landmasses: the Tripura lowlands in the south, the Brahmaputra Valley and the Barak Valley in Assam. Manipur's valley is small, comprising approximately 10% of the state's total land area. The Brahmaputra Valley extends longitudinally from North Lakhimpur to the Dhubri district in Assam, a distance of roughly 730 km. The districts of Cachar, Karimganj, and Hailakandi in South Assam are included in the Barak Valley, which is formed by the river Barak and its tributaries. The Ganga–Brahmaputra plain is extended into the Tripura plain.

The North Eastern Council has estimated a loss of forest coverage to the range of 635 ha in the region in 1993 as compared to 1991 assessment and 316 ha in 1997 as compared to 1995 estimate. The few studies which are available on shifting agriculture in the region are based on macro-level secondary data or carried out only for a part of the area. Hence to get an excellent insight to the problems of shifting agriculture in the region as a whole, it would be necessary to go beyond the overall macro picture presented by aggregate statistics and to carry out investigations at the micro-level. The present study has been undertaken to invigorate micro-level analysis for formulating policy measures to control further degradation of the region on the one hand and to sustain the current level of production on the other.

1.2 Asset Account for Land Use Land Cover (LULC)

Land Use (LU) and Land Cover (LC) are the two important characteristics of classification of land. Land Cover can be defined as observed physical features on the Earth's surface, which transforms to Land Use when a socio-economic function is added to it. Given the fact that increasing anthropogenic activities around the biosphere are causing large-scale alterations of the Earth's land surface, land-use and land-cover change is an important indicator for monitoring environmental changes and a vital input for informed decision making in the context of land management. In India, land-use and land-cover statistics are maintained by National Remote Sensing Centre (NRSC), Department of Space (DOS) through a component on National Land Use and Land Cover (LULC) mapping of the Natural Resources Census (NRC) Project of National Natural Resources Repository (NRR) Program. Thus, LC data indicates bio-physical cover on the earth's surface. On the other hand, LU reflects both the activities undertaken and the institutional arrangements put in place for the purposes of economic production or the maintenance or restoration of environmental functions (EnviStats India, 2020).

The Asset Account for land use land cover for different states of the country has been compiled from the EnviStats India 2018 report published by the Social Statistics Division, Ministry of Statistics and Programme Implementation, Govt. of India. The Asset account for LULC has been broadly categorised into seven groups, viz., (i) Agriculture, (ii) Barren/uncultivable/wastelands, (iii) Builtup, (iv) Forest, (v) Grass/ Grazing, (vi) Snow and glacier and (vii) Wetlands/water bodies. The component of Agriculture is further sub-divided into four categories, viz., (i) Crop land, (ii) Current shifting cultivation, (iii) Plantation and (iv) Fallow. The first three categories consisted of farmland. The classification mainly consisted of opening stock, addition to stock, reduction in stock and closing stock. The details asset account for LULC for the Northeastern states is shown in Table 1.1. It is revealed from the table that 79% of land consisted of shifting cultivation in the opening stock in 2005–06 in Mizoram followed by 65.8% in the state of Nagaland. In the closing stock account in 2011–12 in the state of Mizoram 75.4% land area dominated by current shifting cultivation followed by 66.4% in Nagaland. The other states positions in opening stock are Arunachal Pradesh (27.7%), Manipur (23.3%), Meghalaya (15%), Tripura (3.6%) and Assam (0.5%) respectively in state aggregates of farmland in opening stock category. In closing stock category the states of Mizoram and Nagaland are followed by Arunachal Pradesh (24.7%), Manipur (22.1%), Meghalaya (11.2%), Tripura (5.6%) and Assam (0.5%). There is no shifting cultivation practice in the state of Sikkim. For the NER as whole, shifting cultivation practices covered 10.1 and 9.4% of opening stock and closing stock account in respective farmland. The addition and reduction in stocks of shifting cultivation in asset account indicated that there are increases in shifting cultivation are in the different states of NER except Nagaland and Tripura. On the basis of proportion of area of shifting cultivation in LULC all the states of the NER was categorised into two groups: 20% and above and less than 20% (refer Table 1.2).

Table 1.1 State-wise asset account for shifting cultivation in land use land cover (area in km^2)

State	Opening stock (2005–06)	Addition to stock	Reduction in stock	Closing stock (2011–12)	Opening stock (% of farmland)	Closing stock (% of farmland)
1	2	3	4	5	6	7
Arunachal Pradesh	988.1	462.0	556.4	893.7 (− 94.4)	27.7	24.7
Assam	150.3	82.1	109.5	122.8 (− 27.4)	0.5	0.5
Manipur	490.6	193.5	230.6	453.5 (− 37.1)	23.3	22.1
Meghalaya	322.5	174.0	263.4	233.0 (− 89.4)	15.0	11.2
Mizoram	919.6	693.3	822.4	790.5 (− 129.1)	79.0	75.4
Nagaland	1135.2	951.3	920.8	1165.7 (30.5)	65.8	66.4
Sikkim	0.0	0.0	0.0	0.0	0.0	0.0
Tripura	80.9	119.4	74.8	125.5 (44.6)	3.6	5.6
NER total	4087.1	2675.7	2978.0	3784.8	10.1	9.4

Source Compiled from EnviStats India-2018 (Supplement on Environmental Accounts) Social Statistics Division, Ministry of Statistics and Programme Implementation, Govt. of India
Figures in the parentheses refer to balance between addition and reduction in stock

Table 1.2 State-wise asset account for shifting cultivation in land use land cover

State	Opening stock (% of farmland) (2005–06)	Closing stock (% of farmland) (2011–12)
1	2	3
20% and above category		
Mizoram	79.0	75.4
Nagaland	65.8	66.4
Arunachal Pradesh	27.7	24.7
Manipur	23.3	22.1
Less than 20% category		
Meghalaya	15.0	11.2
Tripura	3.6	5.6
Assam	0.5	0.5
Sikkim	0.0	0.0
NER total	10.1	9.4

Source Compiled from EnviStats India-2018 (Supplement on Environmental Accounts) Social Statistics Division, Ministry of Statistics and Programme Implementation, Govt. of India

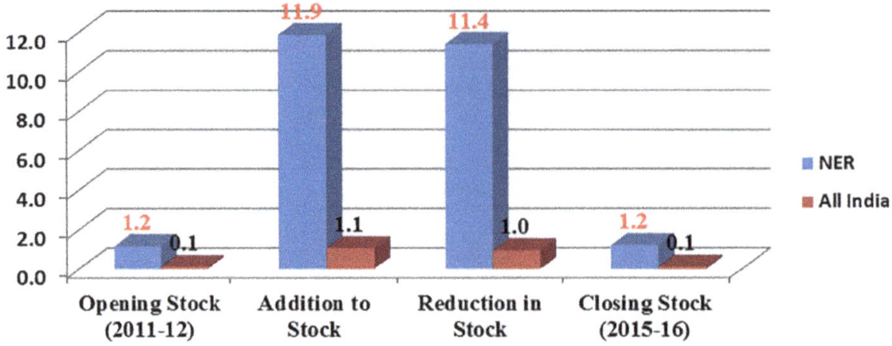

Chart 1.1 Percentage of shifting cultivation to land use land cover 2011–12 to 2015–16. *Source* Compiled from EnviStats India-2020, Vol. I, Social Statistics Division, Ministry of Statistics and Programme Implementation, Govt. of India

A comparison of shifting cultivation practices in the NER with that of All India indicated that dominance of the region in both opening and closing stock of asset distribution of LULC. For instance, in opening stock component (2011–12) 1.2% area belongs to shifting cultivation in NER compared to 0.1% for all India. Similarly in closing stock category for NER shifting cultivation comprised 1.2% of LULC of Agriculture component compared to 0.1% for the country as a whole. For addition and reduction in stock categories 11.9% and 11.4% respectively for NER compared to 1.1% and 1.0% respectively for the country as a whole (refer Chart 1.1).

1.3 The Climate Change, Food Security and Sustainability

The theoretical approach supported by this study was inspired by Connolly-Boutin and Smit (2016). Climate change in the context of the present study indicates various pressures on livelihoods and food security at the community level in the Northeastern region of India and its sustainability. This conceptual framework (Chart 1.2) helps to illustrate the several facets of a community when assessing how vulnerable and adaptable it is to changing circumstances. There are many features of food security and that these are entangled with other characteristics of livelihood and the foundation of natural resources on which communities depend through sustainable practices in management of land resources as in the case of Northeastern region of India.

This conceptual framework offers a broad overview of the links and constraints that shape the dynamics and character of food security in a society. The conditions listed in the framework are probable to differ significantly within a community diverse in character as evidenced in the Northeastern states of India.

According to the findings by Connolly-Boutin and Smit (2016) a community's vulnerability depends on a variety of interconnected biophysical and socio-economic

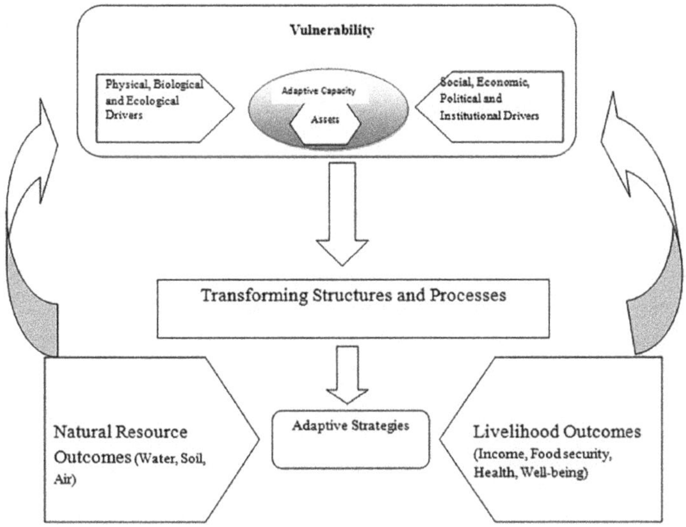

Chart 1.2 Climate change, food security, and sustainability. *Source* Adapted from Connolly-Boutin and Smit (2016)

factors that influence it and mould its capacity for adaptation. Vulnerability is understood to have two interconnected elements, as in the fields of sustainable livelihoods (Chambers, 1989) and sustainability. The first element is related to external risks and pressures that communities are subject to, while the second element is related to coping strategies or adaptive capacity. Climate change and other physical and biological phenomena, such as illness, earthquakes, and soil quality, are examples of biophysical drivers. Institutions policy, economic conditions and demography are examples of socio-economic drivers. Important variables or drivers of vulnerability include cultural norms as relevant for the Northeastern region which is strongly inter-connected among the tribes. Further institutional framework such as forest governance system by the tribal chiefman in the Northeastern region of India play important role for sustainability. The biophysical and socioeconomic drivers overlap, suggesting that their combined effects have a synergistic effect on the community (Ericksen, 2008). This identification of the factors is broadly consistent with concepts in climate change impact assessment, sustainable livelihoods, and food security (Misselhorn, 2005; Scoones, 1998).

The ability of a system to adapt to change is known as adaptive capacity (refer Chart 1.2). Several studies on the vulnerability of climate change have made connections between entitlements or resources or assets used for a livelihood and adaptive capability (Adger, 2006). The Third Assessment Report of the IPCC (2001) defines Climate Change Vulnerability as a function of the character, magnitude, and rate of climate variation to which a system is exposed, its sensitivity, and its adaptive capacity. Adaptive capacity in this study has been defined as the community's capacity to adjust to changing circumstances by using their forest resources which the tribal societies in the Northeastern region depend for their livelihoods. The livelihood resources that people employ in different combinations to build their livelihood adaptation methods are known as assets (Scoones, 2009). The ability of a household member to get employment may be influenced by cultural norms and the climatic conditions. Access to resources may also be impacted by structural and institutional forces. For example, in the Northeastern region many tribal communities have the legal rights to govern and manage the forest resources and distribute among the users on an equitable norms.

Institutional structures and procedures may undergo gradual modifications and changes in response to stressors. The measures people take, either alone or in groups, to adapt to changing circumstances in order to preserve or enhance their well-being are known as adaptation strategies (refer Chart 1.2). As people use their resources to adapt, they are the realisation of adaptive capability (Adger, 2006; Smit & Wandel, 2006). Transformational changes occur when thresholds are crossed in socio-cultural or political economic systems, or when higher-level adjustments are made to policies, programmes, and institutions (Nelson et al., 2007).

Results are a reflection of changes and adjustments. A person's income, health, and food security are among the livelihood outcomes (refer Chart 1.2). According to the theory of sustainable livelihoods, food security is a livelihood outcome (Gladwin et al., 2001; Sutherland et al., 1999). The effects of natural resources, also known as "ecosystem services" (Fisher et al., 2013), can include biodiversity and alterations in the quality of the water, air, or soil.

The fact that the outcomes related to natural resources and livelihoods overlap suggests that they are not independent of one another. For instance, selling cattle as an adaptation strategy to boost revenue could result in the loss of manure and so contribute to the depletion of soil fertility, but it could also reduce overgrazing and improve biodiversity. Moreover, the results can alter the system's susceptibility by providing feedback on its ability to adapt as seen by the diagram's arrows. Positive livelihood outcomes of such higher income that also deplete the basis of natural resources could have unfavourable long-term effects by reducing the amount of physical assets accessible.

The framework shows how well-known concepts from the domains of livelihood vulnerability caused by climate change and the need for sustainable practises for natural resource management can be applied. This framework's primary use is to direct empirical research on the variables and interactions to take into account when determining how climate change is expected to impact food security and livelihoods, specifically in local communities.

Research Questions: In tune with the theoretical framework as discussed above, the present study aims to discourse the major issues concerning climate change and livelihoods of indigeneous tribal societies of India especially existing in the Northeastern region of India. First, the concern is about the associated major factors responsible for climate change in Northeast India? Second, how far rural agricultural livelihoods have been inflated by climate change? Third, what is the status of soil fertility with respect to various crops being grown in different jhum cycles? Fourth, whether the farmers adopted modern technology as mitigation strategies to manage with the climate change?

With the above background relating to the issues identified the study has attempted to the possibilities of sustainable management of CPR based agriculture of these indigenous societies for ensuring food security and sustainability (both economic and ecological). The primary research questions were to make indigenous tribal communities to adapt available agricultural technology to ensure sustainability in the wake of climate changes.

The three critical dimensions, namely, shifting agriculture, the impact of climate changes and livelihood vulnerabilities has been undertaken for an in-depth analysis based on secondary and primary data to assess the effect on bio-resources and on the economy of tribals who are directly dependent on forest resources.

The precise objectives of the study are as follows: The first objective is to examine the intensity of climate change in Northeast India. Second is to identify the major causes of climate change. Third is to investigate the effect of climate change on rural agricultural livelihoods and on soil quality. Fourth is to analyse the technology adoption behaviour of farmers in the changed climatic situation. Fifth is to suggest measures towards achieving sustainable livelihoods of rural masses.

Research Gap: The indigeneous societies are largely affected due to climate change in their shifting agricultural practices which is directly related to man–nature relationships. The tribal communities tried to cope up with the situation using indigeneous practices. The present study has been undertaken a perception-based analysis by evaluating the impact of climate change on agricultural practices taking seven Northeastern states of India under consideration. The previous studies did not consider state-wise analysis on the trend of climatic variables like rainfall, temperature etc. in Northeast India. The impact of climate change on agricultural practices is yet to be explored in these states. Moreover, the existing studies have overlooked the possibility of structural change in trend analysis. The study addresses the livelihoods vulnerabilities and dependency pattern focusing on the importance of community institutions in strengthening sustainable framework for common property resources. Additionally, an investigation has been conducted into the agricultural fields' conditions and the quality of the degraded soil in a few areas of India's northeastern states.

1.4 Research Methodology

The goal of the study was to comprehend how farmers in Northeastern region of India were affected by climate change. The study specifically aimed to determine the extent to which farmers could indicate to the degree of climatic changes that have occurred in the studied locations in recent memory. In addition, the study aimed to determine the extent to which documented climatic changes have affected rural livelihoods and what the future holds for these livelihoods.

The seven Northeastern states are the locations where the study was conducted. To ensure the widest possible spatial coverage, a sample has been chosen from each of the seven states where shifting cultivation is used. Because shifting agriculture is not used in the state of Sikkim, it was not chosen for a field research (refer Table 1.1). However, the state of Sikkim was included in order to understand a comprehensive picture of the NER while evaluating weather variables such as temperature and rainfall, sustainable practices, among other things.

Three methods of gathering data were used in this study in order to triangulate the results. These included observation, in-depth interviews with important informants, and focus group discussions. Focus group discussions (FGDs) were used to bring together different state stakeholders about livelihoods and climate change. A concerted effort was made to make sure that the focus group talks accurately reflect the various parties involved in climate change. Age disparities and gender representation in the focus groups received special attention. The goal of the focus group discussions was to gather information on how the community views climate change and how it has impacted their way of life. Particular inquiries about noteworthy indicators of climate change in nearby communities were made, with an emphasis on temperature changes, precipitation, water source yields, agricultural and animal performance, and ecosystem health in general. It was also mainstreamed to ask questions on how livelihoods had been affected and how communities were adjusting to climate change as it happened.

Comprehensive interviews were carried out with pertinent stakeholders regarding the issue of climate change in Northeast India. Interviews with representatives of Agricultural Extension Services Department were conducted to gain an understanding of their specialised viewpoints on the phenomena of climate change in the research area. Observation was another method of gathering data employed in the study. With this method, the researchers methodically examined the condition of the ecological system with regard to crops, water sources, and grazing areas, among other things. Photographs were taken as part of the observation process.

1.4.1 Sampling Design

It is evident from Tables 1.1 and 1.2 that the states were categorised into two groups' i.e. more than 20% and less than 20% of shifting cultivation area in asset account for LULC. For selection of sample sites the multistage random sample method was used. In the selection of states for conducting sample survey it has been decided to conduct field survey in all the 7 (seven) states where shifting cultivation practices are part of livelihood dependency of the indigenous societies in the NER. Two districts each from all the states were selected at random except for Mizoram and Tripura. In the second stage of selection two villages from one subdivision of each district were selected at random subject to the consideration of availability of shifting cultivation area currently being practiced by the villagers. In the last stage atleast 20 households subject to maximum of 60 households were selected purposively depending upon the practice of shifting cultivation by the household. It is to be noted that in field survey it was found that in all the villages across different states people belonging to indigenous societies live in a cluster household structure in a particular locality. On enquiry by the field personnel it was revealed that due to the threat of wild animals the usual practice of indigenous societies to live in cluster housing complexes instead of scattered locations. For the state of Mizoram and Tripura it was decided to cover all the districts in both these states. This decision was taken due to the fact that Mizoram possesses highest proportion of current shifting cultivation in LULC of closing stock as in 2011–12 and belongs to the category of 20% and above as depicted in Table 1.2. Similarly in the state of Tripura for current shifting cultivation shows large increase i.e. 3.6% in closing stock in 2011–12 as compared to 5.6% in opening stock category in 2005–06 in the less than 20% category. In this category except in Tripura other states experienced reduction in area of current shifting cultivation as compared to addition to stock category and shows decline in proportion in closing stock. Thus, in the two states, viz., Mizoram, where shifting cultivated area is most extensive in the Northeastern region and Tripura, where the reverse trend in terms of increase in area under shifting cultivation was observed, considered for field survey for all the districts in these states. The details of selected sample districts, sub-division and villages state-wise are illustrated in Chap. 4.

References

Agarwal, A. K. (1996). Agro industries for sustainable development of tribal areas in North East India. *Political Economy Journal of India*, July–Dec, *8*(3,4), 127–131.

Adger, W. N. (2006). Vulnerability. *Global Environmental Change, 16*, 268–281. https://doi.org/10.1016/j.gloenvcha.2006.02.006

Adger, W. N., Huq, S., Brown, K., Conway, D., & Hulme, M. (2003). Adaptation to climate change in the developing world. *Progress in Development Studies, 3*(3), 179–195.

Bettinger, P., Clutter, M., Siry, J., Kane, M., & Pait, J. (2009). Broad implications of southern United States pine clonal forestry on planning and management of forests. *International Forestry Review, 11*(3), 331–345.

Brown, O., & Crawford, A. (2008). Assessing the security implications of climate change for West Africa: Country case studies of Ghana and Burkina Faso.

Chambers, R. (1989). Editorial introduction: Vulnerability, coping and policy. *IDS Bulletin, 20*(2), 1–7.

Chambers, R., & Conway, G. (1992). *Sustainable rural livelihoods: Practical concepts for the 21st century*. Institute of Development Studies (UK).

Connolly-Boutin, L., & Smit, B. (2016). Climate change, food security, and livelihoods in sub-Saharan Africa. *Regional Environmental Change, 16*(2), 385–399.

EnviStats India. (2020). Social statistics division, Ministry of Statistics and Programme Implementation, Vol. I, Govt. of India.

Ericksen, P. J. (2008). Conceptualizing food systems for global environmental change research. *Global Environmental Change, 18*(1), 234–245.

Fisher, J. A., Patenaude, G., Meir, P., Nightingale, A. J., Rounsevell, M. D., Williams, M., & Woodhouse, I. H. (2013). Strengthening conceptual foundations: Analysing frameworks for ecosystem services and poverty alleviation research. *Global Environmental Change, 23*(5), 1098–1111.

Gadgil, M., & Guha, R. (1992). *The fissure land: An ecological history of India*. Oxford University Press, Delhi.
Gladwin, C. H., Thomson, A. M., Peterson, J. S., & Anderson, A. S. (2001). Addressing food security in Africa via multiple livelihood strategies of women farmers. *Food Policy, 26*(2), 177–207.
Hussein, K. (2002). *Livelihoods approaches compared*. Department for International Development.
Intergovernmental Panel on Climate Change (IPCC). (2001). *The Third Assessment Report (TAR)*.
Larson, R. P., Byrne, J. M., Johnson, D. L., Kienzie, S. W., & Letts, M. G. (2011). Modelling climate change impacts on spring runoff for the Rocky Mountains of Montana and Alberta II: Runoff change projections using future scenarios. *Canadian Water Resources Journal, 36*(1), 35–52.
Lodha, R. M. (Ed.). (1992). *Environmental ruin: The crisis of survival*. Indus Publishing House.
Mansourian, S., Belokurov, A., & Stephenson, P. J. (2009). The role of forest protected areas in adaptation to climate change. *Unasylva, 60*(231–232), 63–69.
McCarthy, J.J, Canziani, O.F, Leary, N.A and D.J. (Eds.) (2001). Climate Change 2001: Impacts, Adaptation, and Vulnerability: Scenario of the 21st Century. IPCC, Cambridge
Misselhorn, A. A. (2005). What drives food insecurity in southern Africa? A meta-analysis of household economy studies. *Global Environmental Change, 15*(1), 33–43.
Nelson, D. R., Adger, W. N., & Brown, K. (2007). Adaptation to environmental change: Contributions of a resilience framework. *Annual Review of Environment and Resources, 32*.
Ramakrishnan, P. S. (1996). Shifting agriculture and sustainable development. UNESCO and The Parthenon Publishing Group.
Reid, H., & Huq, S. (2005). Climate change-biodiversity and livelihood impacts. In *Tropical forests and adaptation to climate change* (p. 57).
Scoones, I. (1998). *Sustainable rural livelihoods: A framework for analysis* (Vol. 72). Institute of Development Studies.
Scoones, I. (2009). Livelihoods perspectives and rural development. *The Journal of Peasant Studies, 36*(1), 171–196.
Singh, J., Borah, IP., Barua, A., & Barua, K. N. (2003). *Shifting cultivation in North East India–An overview, Rain Forest Research Institute*, http://www.rfri.org/rpap23.htm. Accessed on 5/12/2015.
Smit, B., & Wandel, J. (2006). Adaptation, adaptive capacity and vulnerability. *Global Environmental Change, 16*(3), 282–292.
Sutherland, A. J., Irungu, J. W., Kang'ara, J., Muthamia, J., & Ouma, J. (1999). Household food security in semi-arid Africa—The contribution of participatory adaptive research and development to rural livelihoods in Eastern Kenya. *Food Policy, 24*(4), 363–390.

Chapter 2
Review of Literature

2.1 Literature on the Climate Change and Development Interface

Change in weather over a long period as a result of human or natural influences is defined as climate change. As per Intergovernmental Panel on Climate Change (IPCC) Report, 2007 indicated that climate change would have severe consequences for all living elements. It is one of the most important challenges to long-term sustainability, with negative effects for the atmosphere, human health, food security, economic growth, natural resources, and physical infrastructure. The IPCC has identified five key areas that will be affected by climate change: (i) Water, (ii) Agriculture, (iii) Ecosystem, (iv) Health and (v) Coastlines. Many regions of the world, particularly Sub-Saharan Africa, are facing the possibility of reversed development gains due to the detrimental effects of climate change. The fact that the climate is changing globally is widely acknowledged as a scientific phenomena. Patterns of temperature and precipitation are shifting. Temperatures are rising, precipitation has generally decreased, and rainfall patterns have grown less predictable in the Sub-Saharan region (Kotir, 2011). Evidence shows that the upward trend of the already high temperatures and the reduction of precipitation levels will increasingly result in reduced agricultural production in Sub-Saharan Africa (Dube & Phiri, 2013). The majority of African rural economies are based mostly on agriculture. The other livelihoods that rely on crop production would be negatively impacted by unfavourable trends in the agricultural sector. "Climate change will not be stopped by global action," The Economist (2010) concluded. The world needs to focus more on finding ways to coexist with it. Sub-Saharan African nations have generally suffered from the effects of climate change (Serdeczny et al., 2017). According to Davidson et al. (2003), "it is becoming increasingly clear that climate change can seriously impede the realisation of the development goals." African livelihood protection measures are being undermined by climate change. In order to address development concerns regarding

N. Roy et al., *Climate Change, Shifting Cultivation and Livelihood Vulnerabilities in India*, https://doi.org/10.1007/978-3-031-54927-4_2

livelihoods in Sub-Saharan Africa holistically, relevant development programming will need to increasingly incorporate adaptation to climate change (Ray et al., 2019).

Climate change is a significant and lasting change in the statistical distribution of weather patterns over periods ranging from decades to millions of years. It may be a change in average weather conditions or in the distribution of weather around the average conditions. It is caused by factors that include ocean processes, variations in solar radiation received by the earth, plate tectonics and volcanic eruptions and human-induced alterations of the natural world. The later effects are currently causing global warming, and climate change is often used to describe human-specific impacts. Statistical studies that correlate weather and climate with solar activity have been popular for centuries dating back at least in 1801 (Sah, 2014). The Challenge of handling of statistics is compounded when the weather events being considered over whole seasons. Droughts are usually of most significant interest in respect of the growing seasons in mid-latitudes and terms of the behaviour of the rainy season in more arid parts of the world. The data can vary significantly from year to year, even if it is an average over a few months. What constitutes a few exceptional seasons as opposed to a sustained shift in climate may take many years to establish (Burroughs, 1997).

The concept of externalities has been recognised in neo-classical economics, mainly by Marshall (1910) and Pigou (1932). The classical economists recognised the scarcity of land and other natural resources as a constraint on economic development. It has only been in the last part of twentieth century that economists have made persistent attempts to formally include environmental impacts as a fundamental aspect of all economic activity. Leontief (1970) first tried to consider the environment in general equilibrium and input–output analysis. As one takes a more general equilibrium approach, individual critical relationships between the economy and environment become more evident. First, it is often through the agricultural sectors that the climate is linked most strongly to the overall economy. Agriculture, more so than other sectors, is susceptible to changes in the environment. Climate change, air pollution, water scarcity and loss of plant genetic diversity are examples of environmental stresses that mainly affect agriculture. At the same time, agriculture and forestry sectors are the primary users of land and water resources. As a result, changes in agricultural policies, production and international trade have significant environmental ramifications. Second, evaluating environmental problems requires not just a regional or national perspective but also an international one. Third, there is a vital need for greater understanding of the relationships between agriculture, the environment, poverty alleviation and overall economic growth in developing countries. Agriculture is still a leading sector in developing economies, accounting for significant shares of employment, GDP and foreign exchange. Rural population, especially the poor, depends on access to land and natural resources for their livelihoods and is particularly vulnerable to disruptions in agricultural markets and to environmental degradation. Fourth, a significant research challenge is to make better use of information on environmental relationships in economic modelling.

This includes the problem of choosing the appropriate level of detail and aggregation. Relationships between economic activity and environment are often location-specific. To accurately assess the environmental impacts of economic activity or the economic effects of environmental policies, one needs to account for these spatial differences. Fifth, as one move from partial to general equilibrium analysis, one must confront several second-best problems as emphasised by Mishan (1965) and Ayers and Kneese (1969). The ultimate welfare effects of changes in environmental quality or policies will be sensitive to distortions created by pre-existing trade, tax on agricultural income support policies (Frisvold & Kuhn, 1999).

The impacts of human activities on the world scene and its climate and the continuing effect and magnitude of variations due to natural causes are stressed. Some impact on the human economy—in individual nations and international economy—and on all life on earth today and in the past, have been severe and threaten to remain so. Awareness about climate, its development and changes differ in different communities. It has been presumably, always been high among the more intelligent members of primitive societies living in vulnerable regions, in mountainous districts and near the high latitudes and arid margins of cultivation. This also applies to communities living near exposed coasts, as along the European-Atlantic and the North Sea coasts. It is often minimal among populations living in the sheltered, artificial environment of modern cities. Attitudes also change with time; a single generation or less of immunity from ultimate and other environmental hazards is enough for people to forget about elementary precautions. There have been significant changes in perceptions in the scientific community itself and of the leaders in industry and government. More specifically, in the last four to five years, a tremendous change in attitudes has taken place. The politicians and the public, the news media and the magazines have been gripped by 'greenhouse effect' and the more immediate alarm ozone holes, first reported in 1985 with the implied intensification of the ultraviolet rays in the sun's rays (Lamb, 1993).

Developing nations received special financial and technical help to balance economic development with carbon emission reduction; one way to do this is through the Clean Development Mechanism (CDM) of the Kyoto Protocol. Operating within the CDM is a public–private partnership called the World Bank's Prototype Carbon Fund. UNEP and WTO urged the international community to lower trade barriers to promote commerce in environmental goods and services, hoping to give developing nations more possibilities to adopt sustainable technologies.

2.2 Literature on the Trends of Climatic Variables in India

Several studies have been conducted to understand the dynamics of climate change in India. Many studies observed that India's surface temperature has been rising over the last century (Arora et al., 2005; Dash et al., 2007; Hingane et al., 1985; Rupa Kumar et al., 1994; Srivastava et al., 1992). In one of the early studies on the issue, Hingane et al. (1985), using temperature data at 73 fairly widespread stations for the period 1901–1982, observed that mean annual temperatures in India is rising gradually. The post-monsoon and winter seasons are found to be the major contributors to this warming. In a similar study conducted by Srivastava et al. (1992) covering the study period, 1901–1986 observed that the surface temperatures are showing a decreasing trend in almost all the northern parts of the country and a rising trend in southern regions. However, they found a gradual warming trend across the region. In their research, Rupa Kumar et al. (1994) present linear trend analyses of maximum and minimum temperature data at 121 stations in India from 1901 to 1987. They found that the mean temperature in India is increasing at a steady rate. They went on to claim that the high temperatures are almost entirely responsible for the rise in mean temperatures in India, while the minimum temperatures are virtually trendless, resulting in a rise in the diurnal range of temperatures. Arora et al. (2005) conducted an investigation to detect patterns in temperature time series from 125 stations spread throughout the entirety of India. The yearly average and seasonal temperature trends were examined for monotonic trends using the non-parametric Mann–Kendall test. Three meteorological variables—mean, mean maximum, and mean minimum—were taken into consideration for study on a yearly and seasonal level. The research states that the mean annual temperature, mean maximum temperature, and mean minimum temperature have all increased at rates of 0.42, 0.92, and 0.09 °C (100 years)$^{-1}$, according to the report. In a similar study by Dash et al. (2007) observed that in the last century, the atmospheric surface temperature in India has enhanced by about 1 and 1.1 °C during winter and post-monsoon months respectively. Also, decrease in the minimum temperature during the summer monsoon and its increase during post-monsoon months have created a large difference of about 0.8 °C in the seasonal temperature anomalies which may bring about seasonal asymmetry and hence changes in atmospheric circulation.

The researchers, in general, have observed that the trend and magnitude of warming in India over the years is broadly consistent with the global direction and magnitude. However, the Diurnal Temperature Range (DTR) trends in India are reasonably different from those observed in other parts of the world because of the comparatively large increase in 'maximum daily temperature' over a significant part of India (Rupa Kumar et al., 1994; Srivastava et al., 1992). Similarly, while there is conclusive evidence that the trend of rainfall is decreasing in most of the world, several studies relating to changing pattern of rainfall over India observed that there is no clear trend of increase or decrease in average rainfall over the country (Jain & Kumar, 2012; Kumar et al., 2011; Mondal & Mujumdar, 2015; Vittal et al., 2013). While long-term patterns in monsoon rainfall have not been noted throughout the entirety of India, a number of investigations have discovered noteworthy patterns in the rain (Dash et al., 2007; Guhathakurta et al., 2011; Jain et al., 2013; Kumar et al., 2010; Pingale et al., 2014). While the number of rainy days and average annual precipitation have dropped, some reports indicate that the frequency of intense rainfall events has reduced in many parts of Asia (Dash et al., 2007; Sheikh et al., 2015; Sivakumar & Stefanski, 2010).

In the case of the Northeastern Region of India, the picture is different. Here, the climate is subtropical, and the region is characterised by the wide variation of weather and climate (Choudhury et al., 2012). They observed a substantial rise in temperature, seasonal rainfall, inconsistency in monsoon precipitation, the occurrence of meteorological droughts, wind speed and relative humidity etc. NER is a significant location from a meteorological standpoint because it receives one of the largest amounts of rainfall on Earth, with an average of 2068 mm per year. A few locations in Assam only receive up to 2000 mm of rain annually, making Cherrapunji in Meghalaya the wettest place on Earth with an annual rainfall of almost 12,000 mm.

At four sites in NER, Jhajharia and Singh (2011) showed declining trends in DTR (DTR = Tmax − Tmin) across practically all periods. On the other hand, at the yearly, seasonal (pre-monsoon and monsoon), and monthly (May, June, August, September, and November) durations, the DTR trends were mostly significant. At three sites, significant increases in DTR are noted in October as well as throughout the monsoon and post-monsoon seasons. The T-mean increased significantly in four places during the monsoon and post-monsoon seasons. Additionally, there was evidence of a seasonal cycle since the post-monsoon changes in Tmax and Tmin were greater than the monsoon changes. At the yearly, seasonal, and monthly time-scales (January–March), significant declining trends in the sunlight duration were noted. Using data from eight stations in NER, the authors found that migration increased at three sites and decreased at three others when it came to diurnal temperature ranges. While rising during the monsoon and post-monsoon seasons, temperatures in NER stayed essentially trendless during the winter and pre-monsoon seasons. Jain et al. (2013) looked at trends in temperature and rainfall on a monthly, seasonal, and annual basis at both the subdivision and regional levels. Overall, no discernible pattern has surfaced, either geographically or temporally, despite significant heterogeneity in the size and direction of the trend of rainfall data from one meteorological subdivision to another. The majority of NER states exhibited an increasing trend for all four temperature variables—maximum, minimum, mean, and temperature range—according to trend analysis of the temperature data.

Dikshit and Dikshit (2014) in their study on weather and climate of Northeast India found a significant climatic contrast between the valleys and the mountains. The temperature varied as hugely as in the range of 0–33 °C. The mountains of Arunachal Pradesh experience heavy snowfall. The average rainfall in the region was always seen above 1000 mm. The hilly areas of the Northeast received an average of 2000–3000 mm of rain. In contrast, in Meghalaya, Shillong plateau received highest rains, mean annual rainfall of 11,465 mm. In the region, 90 per cent of the precipitation was expected in the southwest summer monsoon, and June has been identified as the rainiest month by the authors.

Though the above studies have made a significant contribution to understand the trend of climatic variables in NER, yet these studies have focused on the analysis of climatic data for a single station or a group of stations. There is hardly any review on trends of climatic variables taking the state as a unit. In other words, state-wise analysis on the direction of climatic variables is lacking. Secondly, what limited studies have been done those too have overlooked the possibility of structural change in trend analysis.

Most of these studies investigated the trends in annual and seasonal rainfall series on the country scale or in regional scales.

2.3 Literature on the Climate Change and Agricultural Practices in India

For the Indian subcontinent, climate variability and the frequency of extreme events are major concerns. In India, extensive research has been undertaken over the last decade to better understand the extent and magnitude of changes in crop yields as a result of potential climate change. The aim of this review is to (a) analyse the current state of knowledge about the effects of climate change on agricultural production, (b) address the uncertainties and limitations of these studies, and (c) identify potential research needs.

Changes in rising seasonal temperature over time appeared to be the crucial aspect of weather influencing yearly wheat yield variability (Mandal & Nath, 2017). Negative trends in solar radiation and a rise in minimum temperature have resulted in declining trends in potential rice and wheat yields in India's Indo-Gangetic plains (Subash & Mohan, 2012). Both the summer or Kharif growing season (June–October) and the winter or Rabi growing season, which begins after the summer monsoon, display an increasing trend in minimum and maximum temperatures. After 1980, there was also a minor drop in solar radiation during the Rabi and Kharif seasons. These shifts point to a warming trend (Gathala et al., 2013). Since solar radiation is so closely linked to crop growth, any reduction in it would have a direct effect on agricultural productivity (Mall et al., 2018). The subsequent rise in minimum temperatures boosts the crop's maintenance respiration needs, limiting net growth and productivity even further (Birthal et al., 2014; Yadav et al., 2015).

Prasanna (2014) investigated the correlation between Indian Summer Monsoon Rainfall (SMR) and food grain production. Selvaraju (2012) is closely linked to inter-annual fluctuations in SMR and total food grain production anomalies. Kumar et al. (2004) used historical production figures for major crops (rice, wheat, sorghum, groundnut, and sugarcane) as well as aggregate food grain, cereal, pulses, and oilseed production to investigate crop–climate relationships in India. Annual total production in India (except sorghum and sugarcane), output in the monsoon (except sorghum), and output in the post-monsoon (except sorghum and rice) seasons (except sorghum and rice) were all significantly associated with all India summer rainfall.

According to Kumar and Gautam (2014), India will warm over the century, with more seasonal variation in temperature and more heating in the winters than in the summers. They expected a 2.33–4.78 °C rise in average temperature and a doubling of CO_2 concentrations. As a result of these heat waves, the summer monsoon rains will be more variable, and India's agriculture sector will suffer. Mall et al. (2006) found that any rise in mean temperatures from terminal spikelet initiation to anthesis resulted in a decrease in wheat crop. They determined that a 2 °C rise in mean air temperature would reduce rice yield by 0.75 ton/ha in high-yielding areas and 0.06 ton/ha in low-yielding coastal areas. Furthermore, a 0.5 °C rise in winter temperature will shorten the wheat crop by seven days and lower yield by 0.45 ton/ha. A 0.5° rise in winter temperature will result in a 10 per cent reduction in wheat production in Northern India's high-yielding states. Kumar et al. (2014) assessed the vulnerability of wheat production to climate change by using the InfoCrop-WHEAT model. Their study estimated reduction of the wheat production is in the range of 6–23 per cent by 2050 and 15–25 per cent by 2080 due to climate change. The effects are anticipated to be less severe in low-emission scenarios than in high-emission methods.

Vasanta (2013) examined the influence of climate change on India's wheat yields and observed that wheat yields decreased as a result of the negative effects of temperature during grain filling and maturity stages of the growth cycle. According to this findings crop characteristics such as grain filling length sensitivity to temperature play a significant role in assessing the impact of climate change on crop productivity. Lobell et al. (2012) analysed the effect of climate change on wheat production in India in a comprehensive report. In most cases, they discovered that a 2 °C rise in temperature decreased potential yields. However, at low-potential-productivity sites, the magnitude was smaller.

Kumar et al. (2011) examined the climate change impact on the productivity of wheat, rice, maize, potato, coconut, sorghum and mustard crop in the Western Ghats, coastal and Northeastern regions of India. They came to the conclusion that rising temperatures and declining radiation levels pose a serious threat to cereal and crop growth and yields. Changes in variety and agronomy are projected to favour growth and increase crop yields in the future, helping to offset the negative effects of rising temperatures. Kumar (2011) and Mishra et al. (2016) estimated the functional relationship between farm-level net revenue and climate variables, introduced through linear, quadratic, and interaction terms, to understand the climatic sensitivity of Indian agriculture. They discovered that increasing the temperature has major negative effects, while increasing the precipitation in the scenario increases the net revenue. Overall, the negative impacts of temperature change outweigh the slight positive influence of precipitation change.

Ravindranath et al. (2011) attempted to assess the vulnerability of agriculture, water resources and forest sectors to climate variability and climate change in the Northeastern region. Using principal component analysis technique, they observed that majority of the districts in NER are subject to climate-induced vulnerability currently and shortly.

Jamir et al. (2013) use an 'index-based approach' to quantify the climate variability-induced vulnerability of farmers in five villages of Dimapur district, Nagaland in India. Indicators, which are reflective of the exposure, sensitivity and adaptive capacity of the farmers to drought, were quantified based on primary data generated through household surveys and participatory rural appraisal supplemented by secondary data to calculate a composite vulnerability index. The overall results reveal that biophysical characteristics contribute the most to overall vulnerability.

Singh et al. (2017) attempted to study the perception of Adi tribal group in Arunachal Pradesh on climate change and its effect on farming practices. Their research reveals that the perceptions of climate variability and associated livelihood risk are different among male and female farmers belonging to the Adi tribal people. Wealth and gender roles can be used to explain these discrepancies. Men are less conscious of how climate variability affects the activities that typically fall within the purview of women. Contrarily, women believe that almost every aspect of their livelihoods—including cropping, raising animals, fishing, and the fermentation and storage of food—is impacted by climate variability. They are also more aware of how these impacts affect a wide range of their agricultural activities.

2.4 Literature on the Technology in Agriculture

Since the 1960s, scholars have been trying to realise the real meaning of technology and its adoption to agricultural activities. Galbraith (1967) described technology as the systematic application of scientific or other organised knowledge to practical tasks. In one of the first studies on the subject, Knight (1985) proposed that there are four main categories into which contemporary technology might be subdivided: biological technology, which includes new crop varieties, and other technologies that use physical materials. Chemical technology, on the other hand, makes chemical fertilisers and weed and pesticides (fungicides, herbicides, and insecticides) available. Thirdly, mechanical technology, which includes farm machinery and equipment utilised for transporting, weeding, tilling, pest control, and spraying, among other land preparation tasks. Lastly, there is management technology, which encompasses the expertise needed to oversee farming operations and make decisions without explicitly utilising new materials. The technology involves both improved cultivation practices (e.g., proper spacing, proper timing of planting and irrigations) and the use of a variable input (e.g., fertilisers, pesticides) with which the farmers are not familiar (Feder & Slade, 1984). The technology incorporates, among other things, research and development, design, process and production engineering, maintenance, management and entrepreneurship, marketing, investment and finance, human resources, information technology and many others. Technology is always the means of creating new tools serving humans and their environment. Some of the improved technologies are the use of tractors (mechanisation), application of fertilisers and insecticides, adoption of improved spacing, treatment of seed before planting, enhanced storage techniques and a host of others (Oluyole, 2005). Jain et al. (2009) in their study referred to agricultural technologies as all kinds of improved procedures and practices which affect the growth of agricultural output. They included (a) high-yielding varieties of seeds, (b) chemical fertilisers, (c) pesticides, (d) weedicides, and (e) use of machinery in their study. Thus, in agriculture, the term technology generally used to refer to modern technology which centres on technical change through the use of advanced scientific methods of increasing productivity based on mechanical and biological inputs. In agriculture, there are generally two kinds of technology: automated technology which is labour saving and bio-chemical technology which is land saving. Mechanical technology typically involves the use of heavy capital equipment, tractors, thresher etc. Another kind of agricultural technology is biological–chemical technology like improved seeds, fertilisers, pesticides etc. and is labour intensive.

In recent decades, technology transfer has been a major source of concern. The process by which knowledge and technology are dispersed through human activity is known as technology transfer. Transfer of new or improved technology refers to the receivers' or farmers' own practical application of research results or innovations on farms. It implies the trial, evaluation and consequent adoption of the technologies produced (Shand, 1973). Verma (1974) defined the transfer of technology as shifting technology from one person to another and from one place to another place. Webster (1979) defined transfer of agro-technology from the international, national and state centres to the farmers as a vertical process and act as a panacea for social problems like hunger, malnutrition and poverty in the less developed countries.

Transfer of technology generally defined as a process by which the recommended practices produced by research and development agencies are transmitted through extension agents to producers. Therefore, transfer of technology starts after its perfection and ends in its utilisation by the target consumers. Reddy (1996) expressed the view that effective transfer of technology takes place when the maximum number of potential adopters understand, accept and put into practice the central part of an item of technology with the minimum time lag and with the top possible material and financial benefit. Moreover, the process is not only transferring the product but also the use of knowledge and its application. This approach resolved the major analytical problem of differences between technology and knowledge.

Thus, the transfer of technology means the movement of information on technology from the creator to the user through an extension system. In agriculture, it is the farmers who are expected to adopt the new technology in the existing farming system. Farmers are the ultimate users of agricultural technologies developed by the researchers. Any appropriate transfer of technology should include not only shifting of technology from one place to another but also its adoption. There are various stages of technological diffusion, like awareness, experimental, acceptance and adoption. Technology is first developed in the laboratory and then experimented on new farms. If experimentation ends in better results, then the farmers will accept that technique and come forward to adopt it on a massive scale. This is the stage of maturity. The successful spread of new techniques depends on the extension agency. According to Schultz (1956), the major problem that comes in case of technological transformation is how to acquire, adapt and make them acceptable to the farmers for use. He is of the view that three essential steps are involved in the process of the ultimate supply of new inputs to the people who have to use them: (i) Research and development of new inputs, (ii) Distribution of new inputs and (iii) Development of extension services. He also suggested that the development and distribution of new inputs should be taken up by the state or by the non-profit-making agencies.

2.5 Literature on the Adoption of Available Technologies in Agriculture

The rational completion to the process of technology development is its adoption. Unless the technology developed is adopted by the farmers, the gains from the innovations cannot be experienced. The success of any technology depends on its diffusion among the prospective users measured as the level of adoption of that technology. The literature suggests several theoretical or conceptual models on the adoption of new technology by farmers. Rogers (1962) defined adoption as a mental process which goes through a decision-making process beginning with awareness, then the formation of positive or negative attitudes, and finally deciding whether to adopt the technology. The terms "adoption" and "diffusion of technology" refer to two related ideas that describe how innovations are used. Feder et al. (1985) classified adoption as individual adoption and diffusion as aggregate adoption. Adoption means the degree of use of new technology at the individual farmers' level whereas diffusion is the aggregate adoption of new technology within a given region or geographical area (Genius et al., 2014). The different models associated with the adoption of technology consisted of the following.

Imitation Model: Much of the economic research on the adoption of innovation results in the S-shaped diffusion curve. Rogers (1962) found that in most counties, diffusion was an S-shaped function of time. It is because diffusion is a process of imitation whereby a producer comes to know about new technology from others or by observation. Mansfield (1961) considered the case of an industry with identical or homogeneous producers whereby adoption occurs after a 'successful interaction' between a previous buyer and potential adopters. The S-Shaped diffusion curve results from the fact that in the initial period adoptions are very slow. At the early stage of introduction of technology, only a few farmers adopt the technology because of lack of full information about the potential benefits of the technology and also due to the fear of possible risk related with the new technology. After the initial slow start, more agents start adopting new technology which increases the rate of adoption. During this period, the innovation enters the potential market to a large extent. The potential adopters acquire more information about the benefits of the new technology and the degree of risk associated with it. The point at which the diffusion rate peaks and turns sluggish is known as the phase of saturation or inflexion. In most cases, the ceiling is reached before all agents have adopted the new technology (Sunding & Zilberman, 2001). Some may not adopt due to lack of profitability, feasibility, or an even more unique technology may have been adopted instead. There is always a discrepancy among the members of a social system in the manner they respond to innovative idea or practice. While there are still a few members in a social network who are so creative that they adopt an innovation almost immediately after they come to know about it, the majority take a long time before accepting the new idea or practice. It is the first few adopters of innovation who influence the other members of a community to adopt the innovation.

Griliches (1957) and Mansfield (1961) demonstrated empirically that profitability could hasten and expand the diffusion process. They found that factors such as the profitability of the technology, the amount of money needed to implement it, the degree of uncertainty, and its accessibility all affect how quickly new technologies are adopted.

However, the imitation model is a macro-economic diffusion model dealing with the adoption of new technology at the aggregate level and does not explicitly indicate the decision making by individual firms for adopting new technology. Moreover, the imitation model assumed that potential adopters are homogeneous, whereas heterogeneity can play a significant role in an individual's decision to adopt. To deal with the problems of the microeconomic decision-making process and heterogeneous adopters David (1975) introduced another model of adoption and diffusion namely, threshold mode which is a micro-economic model. This model was used to explain the adoption of grain harvesting machinery in the United States in the 19th century.

Threshold model: The threshold model introduced three elements in the adoption and diffusion model (a) micro-level maximising or satisfying behaviour such as profit maximization, utility maximization etc., (b) heterogeneous characteristics of individual agents such as farm size, land quality, level of risk aversion, human capital and (c) dynamic process like learning by doing or learning by using.

The time of adoption of different individual units may differ depending on the threshold level of heterogeneity. According to David (1975), the main source of heterogeneity among the farmers is the farm size. The adoption of new technology requires a fixed cost which varies over time, and the diffusion process occurs as the fixed cost of the modern technology declines overtime or the variable cost differential between the two technologies increases over time. 'Learning by doing' may reduce fixed costs of new technology through knowledge accumulation and 'Learning by using' increases the adopters' benefit from using the technology over time. The shape of the diffusion curve depends on the dynamics of farm size and the shape of farm size distribution. When the new innovation is introduced, only farms with large size found easier to adopt it. The critical size declines over time and this change triggers more adoption. Marginal diffusion increases during the initial period, and then it falls, thus leading to an S-shaped diffusion curve. The threshold model also applies in other cases where heterogeneity results from differences in land quality or human capital (Caswell & Zilberman, 1986).

Wozniak (1984) argued that uncertainty and fixed costs of adoption of new technology create incentives for farmers to resist early acceptance. Differences in the capacity to overcome that resistance are attributable to differences in education and experience. Therefore, more educated and experienced farmers are more likely to be early adopters than other farmers. Producers who are better informed about innovation are more likely to be innovative than producers with less information. Agricultural extension service and private agricultural firms are the two most relevant sources of technical information on new farming products and procedures. The model was used by Feder and Slade (1984) to analyse the situation where farmers are deciding how much land to set aside for new technologies. The main causes of farm-level variation in this instance, according to the authors, were farmers' earlier evaluations of the benefits of the new technology and knowledge gained from early adopters' accumulated information. Using firm size as the source of variability, Stoneman and Ireland (1983) used the threshold model to a generic scenario of technological adoption. They discovered that larger companies were the first to use the cost-cutting technology, and that the factors propelling its continuing distribution were the technology's falling price and the rising cost of competing for substitute inputs.

Thus, the imitation model focuses on the role of social networks, knowledge transfer, and "imitation" in this process and aims to provide a comprehensive macroeconomic portrayal of the diffusion process. This is especially important in situations when individual producers lack the capacity to regularly evaluate emerging technologies and may not be aware of the range of technologies available. Therefore, the model suggests that imitation plays a role in the information transfer process that results in the spread of new technologies. The diffusion process involving diverse individual actors is not taken into account in this model. In contrast, the threshold model requires each person to evaluate the new technology based on the source of heterogeneity, the individual-level maximisation problem, and particular dynamic processes at the precise moment when the economic conditions are favourable for adoption. Therefore, a mechanism that encourages farmers to embrace the technology at the right moment is needed for the threshold model. This does not mean, however, that the threshold model does not include imitation; on the contrary, it implicitly incorporates the process of reproduction. Adoption relationships that are dynamic and measurable are produced by both models. The diffusion model states that factors such as farm size, risk, human capital, and learning can affect how quickly new technologies are adopted.

The combination of these impacts will assess the overall influence of climate change on agriculture. Assessment of the effect of global climate change on agriculture will assist in accurately forecasting and adapting farming to increase agricultural production. At the same time agriculture has been shown to produce significant effects on climate change, primarily through the production and release of greenhouse gases. Land use change such as deforestation and desertification together with the use of fossil fuels are the major anthropogenic sources of carbon dioxide, agriculture itself is the major contributor to increasing methane and nitrous oxide concentrations in Earth's atmosphere (Sah, 2014).

In the Second Assessment Report (Dec. 1995) the UNIPCC (United Nations Intergovernmental Panel on Climate Change) concluded that 'the balance of evidence suggests a discernable human influence on global climate'. A change in climatic conditions will affect agricultural production systems the world over (Bazzaze & Sombroek, 1996). Climate change poses a danger since it requires the involvement of many processes with various variables that must be taken into account collectively. Agriculture is one of the processes and climate is the other. If these processes are handled separately, the result would be a too fragmented approach. This is a more global problem and human actions are now thought to have the ability to influence climate, one of the environment's components. Agriculture, which provides all food for humans and domestic animals, is affected by the climate. It is also important to note that not only the environment is changing, but human communities and agriculture are creating their own patterns and constraints. This transition need to be taken into consideration by climate change effect studies. There is also the concern of long-term viability in agriculture and rural development. In the next 50 years, how will the relationship between environmental capital and demography change? Will it be possible to increase food production at the same time as protecting environmental resources such as soils and biodiversity? Infact improved nutrition and food security are two parts of a spiral that also includes population pressures and rural poverty.

There are concerns about breaking the cycle of poverty in many developing countries as climatic conditions change. In general, global food production has outpaced human population growth. However, there are major variations between continents. For example, in Africa, local cereal production cannot keep pace with population growth, and root and tuber crop production is rising faster than that of nutritionally more valuable grains. Arable land expansion, on the other hand, lags behind population growth, suggesting some intensification of development. In many Asian countries, the upper limit of arable land has been reached, resulting in very high cropping intensities and irrigation playing a dominant role. In Latin America, increasing arable land comes at a high environmental cost due mainly to deforestation, which may have direct implications for climate change (Sombroek & Gommens, 1996).

There is a vast difference exist among developing nations in respect of agricultural development and population increase. In developed countries, the differences tend to level out. Agriculture is threatened by global climate change. The majority of processes and interactions between agriculture and environment are well known, even if they are not always obvious. It is also clear that the connection between climate change and agriculture is still concerned with uncertainty. On the other hand, climate change is an issue in which every human being on the planet is a potential stakeholder. Since greenhouse gas emissions at any location or point in time will affect the lives of everyone in the world, at all future times, it is the ultimate environmental externality. Therefore, equity and ethical considerations have an essential role to play in determining a climate change response strategy that will prevent a catastrophic outcome. Ongoing climate change negotiations among nations are based on several factors, including risk, fear and the precautionary principle. This approach favours greenhouse gas mitigation efforts via a global market that is already well developed and eager for profits derived from trading in emission reductions, rather than through more deep rooted alterations in social attitudes (Pinguelli-Rosa & Munasinghe, 2002). Traditional ecological knowledge, which covers all topics related to ecology and natural resource management, has become more important in light of global change and concerns for biodiversity. These insights yield three types of benefits: (i) economic—traditional crop varieties, lesser-known edible plants and animals, medicinal plants, etc. harvested from the wild; (ii) ecological or social manipulation of biodiversity for adapting to environmental uncertainties and climate change; (iii) systems of ethical, cultural, spiritual, and religious belief focused on the idea of holy groves, holy landscapes, and sacred creatures (Ramakrishnan, 2001).

Summing Up

The above studies have essential contributions to understand the impact of climate change on agricultural production and productivities in India and the role of technological innovations. It has been widely studies about the adaptation behaviours which vary across different societies in the world. However, there are minimal studies that have investigated the effects of climate change on farming methods which are practiced by the indigenous societies in the Northeast India. The studies that have been conducted by a few researchers in Northeastern states of India have limited their investigation to qualitative analysis and limited to particular state.

References

Arora, M., Goel, N. K., & Singh, P. (2005). Evaluation of temperature trends over India. *Hydrological Sciences Journal, 50*(1), 81–93.

Ayers, R., & Kneese, A. (1969). Production, consumption and externalities. *American Economic Review, 59*(3), 282–297.

Bazzaze, F., & Sombroek, W. G. (1996). *Global climate change and agricultural production*. Food and Agriculture Organisation of the United Nations, Wiley. ISBN: 0-471-95763-1.

Birthal, P. S., Khan, T., Negi, D. S., & Agarwal, S. (2014). Impact of climate change on yields of major food crops in India: Implications for food security. *Agricultural Economics Research Review, 27*(2), 145–155.

Burroughs, J. W. (1997). *Does the weather really matter? The social implications of climate change*. Cambridge University Press. ISBN: 0521-561-264.

Caswell, M. F., & Zilberman, D. (1986). The effects of well depth and land quality on the choice of irrigation technology. *American Journal of Agricultural Economics, 68*(4), 798–811.

Choudhury, B. U., Das, A., Ngachan, S. V., Slong, A., Bordoloi, L. J., & Chowdhury, P. (2012). Trend analysis of long term weather variables in mid altitude Meghalaya, North-East India. *Journal of Agricultural Physics, 12*(1), 12–22.

Dash, S. K., Jenamani, R. K., Kalsi, S. R., & Panda, S. K. (2007). Some evidence of climate change in twentieth-century India. *Climatic Change, 85*, 299–321.

David, P. A. (1975). *Technical choice, innovation and economic growth*. Cambridge University Press.

Davidson, D. J., Williamson, T., & Parkins, J. R. (2003). Understanding climate change risk and vulnerability in northern forest-based communities. *Canadian Journal of Forest Research, 33*(11), 2252–2261.

Dikshit, K. R., & Dikshit, J. K. (2014). Weather and climate of north-east India. In *North-east India: Land, people and economy* (pp. 149–173). Springer, Dordrecht.

Dube, T., & Phiri, K. (2013). Rural livelihoods under stress: The impact of climate change on livelihoods in South Western Zimbabwe. *American International Journal of Contemporary Research, 3*(5), 11–25.

Feder, G., & Slade, R. (1984). The acquisition of information and the adoption of new technology. *American Journal of Agricultural Economics, 66*(3), 312–320.

Feder, G., Richard, E., & Zilberman, D. (1985). Adoption of agricultural innovations in developing countries: A survey. *Economic Development and Cultural Change. 33*(2), 255–98. https://doi.org/10.1086/451461.

Frisvold, G., & Kuhn, B. (1999). *Global environmental change and agriculture: Assessing the impacts*. Edward Elgar Publishing Ltd. ISBN: 185-8986-311.

Galbraith, J. K. (1967). Review of a review. *The Public Interest, 9*, 109.

Gathala, M. K., Kumar, V., Sharma, P. C., Saharawat, Y. S., Jat, H. S., Singh, M., Kumar, A., Jat, M. L., Humphreys, E., Sharma, D. K., Sharma, S., & Ladha, J. K. (2013). Optimizing intensive cereal-based cropping systems addressing current and future drivers of agricultural change in the northwestern Indo-Gangetic Plains of India. *Agriculture, Ecosystems & Environment, 177*, 85–97.

Genius, M., Koundouri, P., Nauges, C., & Tzouvelekas, V. (2014). Information transmission in irrigation technology adoption and diffusion: Social learning, extension services, and spatial effects. *American Journal of Agricultural Economics, 96*(1), 328–344.

Griliches, Z. (1957). Hybrid corn: An exploration in the economics of technological change. *Econometrica, 25*, 501–522.

Guhathakurta, P., Sreejith, O. P., & Menon, P. A. (2011). Impact of climate change on extreme rainfall events and flood risk in India. *Journal of Earth System Science, 120*(3), 359.

Hingane, L. S., Rupa Kumar, K., & Ramana Murthy, Bh. V. (1985). Long-term trends of surface air temperature in India. *Journal of Climatology, 5*, 521–528.

Jain, R., Arora, A., & Raju, S. S. (2009). A Novel Adoption index of selected agricultural technologies: Linkages with infrastructure and productivity. *Agricultural Economics Research Review*, Agricultural Economics Research Association (India), vol. 22(1), January.

Jain, S. K., & Kumar, V. (2012). Trend analysis of rainfall and temperature data for India. *Current Science*, 37–49.

Jain, S. K., Kumar, V., & Saharia, M. (2013). Analysis of rainfall and temperature trends in northeast India. *International Journal of Climatology, 33*(4), 968–978.

Jamir, C., Sharma, N., Sengupta, A., & Ravindranath, N. H. (2013). Farmers' vulnerability to climate variability in Dimapur district of Nagaland, India. *Regional Environmental Change, 13*(1), 153–164.

Jhajharia, D., & Sing, V. P. (2011). Trends in temperature, diurnal temperature range and sunshine duration in Northeast India, July. *International Journal of Climatology, 31*(9), 1353–1367. https://doi.org/10.1002/joc.2164.

Knight, K. E. (1985). A functional and structural measurement of technology. *Technological Forecasting and Social Change, 27*(2–3), 107–127.

Kotir, J. H. (2011). Climate change and variability in Sub-Saharan Africa: A review of current and future trends and impacts on agriculture and food security. *Environment, Development and Sustainability, 13*(3), 587–605.

Kumar, K, K., Rupa K. K., Ashrit, R. G., Deshpande, N. R., & Hansen, J. W. (2004). Climate impacts on Indian agriculture. *International Journal Of Climatology*, 24, 1375–1393, Wiley InterScience (https://www.interscience.wiley.com). https://doi.org/10.1002/joc.1081.

Kumar, K. K. (2011). Climate sensitivity of Indian agriculture: Do spatial effects matter? *Cambridge Journal of Regions, Economy and Society, 4*(2), 221–235.

Kumar, R., & Gautam, H. R. (2014). Climate change and its impact on agricultural productivity in India. *Journal of Climatology & Weather Forecasting*.

Kumar, S. N., Aggarwal, P. K., Rani, D. S., Saxena, R., Chauhan, N., & Jain, S. (2014). Vulnerability of wheat production to climate change in India. *Climate Research, 59*(3), 173–187.

Kumar, S. N., Aggarwal, P. K., Rani, S., Jain, S., Saxena, R., & Chauhan, N. (2011). Impact of climate change on crop productivity in Western Ghats, coastal and northeastern regions of India. *Current Science*, 332–341.

Kumar, V., Jain, S. K., & Singh, Y. (2010). Analysis of long-term rainfall trends in India. *Hydrological Sciences Journal, 55*(4), 484–496. https://doi.org/10.1080/02626667.2010.481373.

Lamb, H. H. (1993). Climate change: The development of knowledge, problems of assessing trends, causes and human impact. In M. Lal (Ed.), *Global warming: Concern for tomorrow*. Tata McGraw-Hill Publishing Company Ltd. ISBN: 0-07-460-211-X.

Leontief, W. (1970). Environmental repercussions and the economic structure: An input-output approach. *Review of Economics and Statistics*.

Lobell, D. B., Sibley, A., & Ortiz-Monasterio, J. I. (2012). Extreme heat effects on wheat senescence in India. *Nature Climate Change, 2*(3), 186–189.

Mall, R. K., Singh, N., Singh, K. K., Sonkar, G., & Gupta, A. (2018). Evaluating the performance of RegCM4.0 climate model for climate change impact assessment on wheat and rice crop in diverse agro-climatic zones of Uttar Pradesh, India. *Climatic Change, 149*(3–4), 503–515.

Mall, R. K., Singh, R., Gupta, A., Srinivasan, G., & Rathore, L. S. (2006). Impact of climate change on Indian agriculture: A review. *Climatic Change, 78*(2–4), 445–478.

Mandal, R., & Nath, H. K. (2017). Climate change and Indian agriculture: Impacts on crop yield. *Food Industry eJournal.*

Mansfield, E. (1961). Technical change and the rate of imitation. *Econometrica, 29*, 741–765.

Marshall, A. (1910). *Principles of economics.* Macmillan.

Mishan, E. J. (1965). Reflections on recent developments in the concept of external effects. *Canadian Journal of Economics and Political Science, 31.*

Mishra, D., Sahu, N. C., & Sahoo, D. (2016). Impact of climate change on agricultural production of Odisha (India): A Ricardian analysis. *Regional Environmental Change, 16*(2), 575–584.

Mondal, A., & Mujumdar, P. P. (2015). Modeling non-stationarity in intensity, duration and frequency of extreme rainfall over India. *Journal of Hydrology, 521*, 217–231.

Oluyole, K. A. (2005). Evaluation of the economics of post harvest processing of cocoa in Cross River State, Nigeria. *Journal of Agriculture, Forestry and the Social Sciences, 3*(2), 58–64.

Pigou, A. C. (1932). *The economics of welfare.* Macmillan.

Pingale, S. M., Khare, D., Jat, M. K., & Adamowski, J. (2014). Spatial and temporal trends of mean and extreme rainfall and temperature for the 33 urban centers of the arid and semi-arid state of Rajasthan, India. *Atmospheric Research, 138*, 73–90.

Pinguelli-Rosa, L., & Munasinghe, M. (2002). *Ethics, equity and international negotiations on climate change.* Edward Elgar. ISBN: 1843760-487.

Prasanna, V. (2014). Impact of monsoon rainfall on the total foodgrain yield over India. *Journal of Earth System Science, 123*(5), 1129–1145.

Ramakrishnan, P. S. (2001). *Ecology and sustainable development.* National Book Trust, India.

Ravindranath, N. H., Rao, S., Sharma, N., Nair, M., Gopalakrishnan, R., Rao, A., Malaviya, S., Tiwari, R., Sagadevan, A., Munsi, M., Krishna, N., & Govindasamy, B. (2011). Climate change vulnerability profiles for North East India. *Current Science, 101*(3). Available at SSRN: https://ssrn.com/abstract=2140671

Ray, D. K., West, P. C., Clark, M., Gerber, J. S., Prishchepov, A. V., & Chatterjee, S. (2019). Climate change has likely already affected global food production. *PLoS ONE, 14*(5), e0217148.

Reddy, A. (1996). *A macro perspective on technology transfer.* Bloomsbury Publishing USA.

Rogers, E. M. (1962). *Diffusion of innovations.* Free Press of Glencoe.

Rupa Kumar, K., Krishna Kumar, K., & Pant, G. B. (1994). Diurnal asymmetry of surface temperature trends over India. *Geophysical Research Letters, 21*, 677–680.

Sah, S. (2014). *Climate change and food security.* KK Publications. ISBN: 978-81-7844-179-5.

Schultz, T. W. (1956). Reflections on agricultural production, output and supply. *Journal of Farm Economics, 38*(3), 748–762.

Selvaraju, R. (2012). Climate risk assessment and management in agriculture. In *Building resilience for adaptation to climate change in the agriculture sector* (Vol. 23, p. 71).

Serdeczny, O., Adams, S., Baarsch, F., Coumou, D., Robinson, A., Hare, W., Schaeffer, M., Perrette, M., & Reinhardt, J. (2017). Climate change impacts in Sub-Saharan Africa: From physical changes to their social repercussions. *Regional Environmental Change, 17*(6), 1585–1600.

Shand, R. T. (1973). (Ed). *Technical change in Asian agriculture.* Canberra Australian National University Press.

Sheikh, M. M., Manzoor, N., Ashraf, J., Adnan, M., Collins, D., Hameed, S., Manton, M. J., Ahmed, A. U., Baidya, S. K., Borgaonkar, H. P., Islam, N., Jayasinghearachchi, D., Kothawale, D. R., Premalal, K. H. M. S., Revadekar, J. V., & Shrestha, M. L. (2015). Trends in extreme daily rainfall and temperature indices over South Asia. *International Journal of Climatology, 35*(7), 1625–1637.

Singh, R. K., Zander, K. K., Kumar, S., Singh, A., Sheoran, P., Kumar, A., Hussain, S. M., Riba, T., Rallen, O., Lego, Y. L., Padung, E., & Garnett, S. T. (2017). Perceptions of climate variability and

livelihood adaptations relating to gender and wealth among the Adi community of the Eastern Indian Himalayas. *Applied Geography, 86*, 41–52.
Sivakumar, M. V., & Stefanski, R. (2010). Climate change in South Asia. In *Climate change and food security in South Asia* (pp. 13–30). Springer, Dordrecht.
Sombroek, W. G., & Gommens, R. (1996). The climate change—Agriculture conundrum. In F. Bazzaze & W. Sombroek (Eds.), *Global climate change and agricultural production*. Food and Agriculture Organisation of the United Nations, Wiley. ISBN: 0-471-95763-1.
Srivastava, H. N., Dewan, B. N., Dikshit, S. K., Rao, G. S. P., Singh, S. S., & Rao, R. (1992). Decadal trends in climate over India. *Mausam, 43*, 7–20.
Stoneman, P., & Ireland, N. J. (1983). The role of supply factors in the diffusion of new process technology. *The Economic Journal, 93*(Supplement), 66–78.
Subash, N., & Mohan, H. R. (2012). Evaluation of the impact of climatic trends and variability in rice–wheat system productivity using Cropping System Model DSSAT over the Indo-Gangetic Plains of India. *Agricultural and Forest Meteorology, 164*, 71–81.
Sunding, D., & Zilberman, D. (2001). The agricultural innovation process: Research and technology adoption in a changing agricultural sector. In B. Gardner & G. Rausser (Eds.), *Handbook of agricultural economics* (Vol. 1A, pp. 207–261).
The Economist. (2010). Facing the consequences, Nov 25th, https://www.economist.com/briefing/2010/11/25/facing-the-consequences.
Vasanta, K. (2013). Impact of climate change on wheat and rice production: An analysis. *Economic Affairs, 58*(2), 89–95.
Verma, O. S. (1974). Problems of transfer of technology and distance as a measure of the system. In S. Daulat & J. S. Pandey (Eds.), *Transfer technology: Systems and constraints*, U.P. Society of Extension Education and Rural Development, India.
Vittal, H., Karmakar, S., & Ghosh, S. (2013). Diametric changes in trends and patterns of extreme rainfall over India from pre-1950 to post-1950. *Geophysical Research Letters, 40*(12), 3253–3258.
Webster, P. R. (1979). Conducting educational research. *Journal of Research in Music Education*. https://doi.org/10.2307/3344713.
Wozniak, G. D. (1984). The adoption of interrelated innovations: A human capital approach. *Review of Economics and Statistics, 66*(1), 70–79.
Yadav, M. K., Singh, R. S., Singh, K. K., Mall, R. K., Patel, C. B., Yadav, S. K., & Singh, M. K. (2015). Assessment of climate change impact on productivity of different cereal crops in Varanasi, India. *Journal of Agrometeorology, 17*(2), 179–184.

Chapter 3
An Overview on Northeastern States of India

3.1 Locations and Boundary

The Northeastern region of India is surrounded by international borders in the north by China; in the south west by Bangladesh; in the north west by Bhutan and in the east by Myanmar. The region is situated between longitude 89.46 and 97.30° East and latitude 21.57 and 29.30° north covering an area of 2.55 lakh km^2 area is surrounded by plains on both sides of the Brahmaputra river and the Himalayan range. This region's flora and fauna are diverse and plentiful. The region accounts for 7.8% of the country's total land area. The Himalayan mountain ranges have defended the region's northern frontier. Plains slightly above sea level and mountains above the snow line make up the terrain. Since nearly 90% of the region's boundaries shape India's international borders, the region is strategically important to the country (Agarwal, 1996).

3.1.1 Physiographical Features

About 70% of the Northeastern region of India is hilly, with different states having different topographies. Most of Arunachal Pradesh, Mizoram, Nagaland, Meghalaya, half of Tripura, one-fifth of Assam, and nine-tenths of Manipur are mountainous and hilly regions. The plain parts of the region consist mostly of various land masses, like the Tripura plains in the south and the Brahmaputra and Barak valleys in Assam. The narrow valley in Manipur makes up only 10% of the state's total land area. Approximately 730 km stretched the Brahmaputra Valley in Assam, from North Lakhimpur to the Dhubri region. The Barak Valley, which is created by the river Barak and its tributaries, runs through the south Assam districts of Cachar, Karimganj, and Hailakandi. The Ganga–Brahmaputra plain is extended by Tripura. The hills' topography is typically rugged, and vast swaths of land are inaccessible. The

N. Roy et al., *Climate Change, Shifting Cultivation and Livelihood Vulnerabilities in India*, https://doi.org/10.1007/978-3-031-54927-4_3

area is comprised an abundance of ecosystems characterised by diverse biota with a high degree of endemism at the confluence of the Indo-Malayan and Palearctic biogeographical realms. More than 200 out of 450 Indian indigenous tribes live in this area, whose culture and customs play an important role in providing lessons on biodiversity.

3.1.2 Climatic Conditions

The climate of the Northeastern region of India is sub-tropical, warm and humid. In this region, the rainy season often starts in March and lasts until the middle of October. The yearly rainfall average in the area varies greatly. The most rainfall, approximately 1080 cm, is experienced annually in the Khasi and Jaintia hills of Meghalaya, specifically in Cherrapunjee and Mawsynram. It is marginally lower in the Nagaon district of Assam, which is under a rain shadow. The four monsoon months of June through September account for around two thirds of the total yearly precipitation.

The quick variations in topography have brought about fast changes in the climate. In January, the daily temperature is typically between 100 and 150 °C in other areas of the region, although it is typically about 15 °C in the plains of Brahmaputra, the Barak Valley, Tripura, and the western section of the Mizo hills. With the exception of Shillong and the southeast part of the Mizo hills, the mean temperature rises in April and falls in July, ranging from 25 to 27.5 °C. In October, the average daily temperature in the hilly regions is 20–25 °C, whereas it is approximately 25 °C in Brahmaputra and Barak Valley, Tripura, and the western part of the Mizo hills. Furthermore, Arunachal Pradesh experiences the lowest temperatures below freezing point in the upper Himalayas.

3.2 Demographical Features

As per the 2011 census, the Northeastern region of India had 4.5 crore people living there. In the region, roughly 81.5% and 18.5% of people, respectively, reside in rural and urban areas, compared to 69% and 31% in the country as a whole. Urbanisation is therefore happening at a slower rate than at the national level. Currently, the region's population makes up roughly 3.72% of the India's total population. Among the seven states in the region the density of population is higher in Assam (398) and lowest in Arunachal Pradesh (17). The densities in other states are Tripura (350), Manipur (115), Meghalaya (132), Nagaland (119) and Mizoram (52). The region is marked by uneven spatial distribution of population among the constituent states, the primary reason being that the plains and the valleys offer more congenial conditions for absorption of population than the hills and difficult terrains. Apart from Assam and Tripura the Northeastern states are mostly inhabited by tribes having unique social and cultural practices. About 75% of the Northeastern region is covered by hilly terrain where indigenous tribal communities mainly live. Overall, tribal people account for over 56% of the total population of the region (Govt. of India Census, 2011). However, in Meghalaya, Mizoram and Nagaland scheduled tribes comprise more than 80% of the population, the proportion for Arunachal Pradesh is 68%. That is what gives these states a predominantly tribal character. Compared to the average for all of India, the population growth rate in all of the NER states is significantly greater. The region's population has a higher overall literacy rate than all India average. Mizoram has the second highest literacy rate in the country, while Assam, Arunachal Pradesh and Meghalaya are below the country average. The important demographic characteristic of the North Eastern states is the country's higher concentration of scheduled tribes. The percentage of people living below the poverty line is also higher in all the North Eastern states than the country as a whole (refer Table 3.1). While there has been a significant population influx in some states, such as Assam, Tripura, and to some extent Meghalaya and Mizoram, the natural growth rate in other states is significantly higher than the national average for India. Cultivators and farm workers made up more than half of the working population, comparable to the rest of India.

The tribal population of the Northeastern region is classified on the basis of territory, language, occupation and cultural contacts and as such widely varies from state

Table 3.1 Demographic indicators of Northeastern states of India as per 2011 census

States	Population (million)	Rural population (%)	Scheduled tribe population (%)	Literacy rate (%)	Working population (percentage of total population)	Poverty ratio (%)
1	2	3	4	5	8	10
Arunachal Pradesh	1.4	77.1	68.8	67.0	42.47	34.67
Assam	31.2	85.9	12.4	73.2	38.36	31.98
Manipur	2.9	70.8	40.9	79.9	45.68	36.89
Meghalaya	3.0	79.9	86.1	75.5	39.96	11.87
Mizoram	1.1	47.9	94.4	91.6	49.24	20.40
Nagaland	2.0	71.1	86.5	80.1	44.36	18.88
Sikkim	0.6	74.8	33.8	82.2	50.47	8.19
Tripura	3.7	73.8	31.8	87.8	40.00	14.05
North East India	45.8	81.6	27.7 (63.18[a])	79.6	39.95	22.12
All India	1210.9	68.9	8.6	74.0	39.80	21.92

Source (1) Registrar General of India 2011, (2) Agricultural Statistics at a Glance, 2014, Ministry of Agriculture, Govt. of India and (3) Planning Commission, Government of India

[a] Figure excluding state of Assam, which is not a tribal dominated state

to state and in some cases intra state variations are also visible. They are dominant ethnic groups in the hill states of Mizoram, Meghalaya, Nagaland, Arunachal Pradesh and in the hill districts of Assam, Manipur and Tripura classified administratively by Autonomous District Council as per the provisions of the sixth schedule of the constitution of India. But Northeastern India's uniqueness goes beyond its physical characteristics; it also has a unique people composition and cultural pattern. Indigenous tribal societies have lived in Northeastern region of India from ancient times. The Indo Mongoloids' dominance in terms of racial and cultural traits is too evident, even though scholars have identified Austric and Dravidian substrata in the region's racial and cultural makeup and although waves of Aryan culture have swept over parts of the region, mainly in the valleys of Assam, Manipur, and Tripura (Medhi et al., 2009).

3.3 Land Use Pattern and Its Dimensions

According to the report of North Eastern Council (Basic Statistics of Northeastern Region, 2015) on land utilisation, out of the 26.21 lakh hectare of the total geographical area of the region, 23.21 lakh hectare (88.55%) is the reporting area for various types of land uses. Of the total reporting area, net area sown forms only 10.25% whereas at the national level it is 46.28%. A straight comparison of this nature between the NER and the country as a whole is perhaps not proper in respect of land utilisation because of the basic differences in geophysical characteristics. For example, 44.20% of the region's reporting area in 2011 was under forests as against 21.29% in the country, which cannot be put to cultivation. The percentage of cultivable area over the geographical area in most of the Northeastern states (26.36%) is less than the national average (32%). The percentage of cultivable area over the geographical area is highest in Assam (53.92%) and lowest in Arunachal Pradesh (4.61%). The percentage utilisation of cultivable area in the region (87.20%) is more than the national average (46.28%) and mono cropping is the predominant method of cultivation. In the absence of intensive multi-cropping, little or no surplus is generated in the agricultural sector. The utilisation is lowest in Nagaland (43.46%) and highest in Manipur (100%) (refer Table 3.2).

3.4 Agricultural Land and Land Holding Pattern

The hills constituted about 70% of the total land area and the low percentage of cultivable land and sparse population in the hill states limits the choices for land based activities in the hills. Due to the difficulty of storing water in view of hill slopes, farmers have been forced to rely heavily on jhum cultivation (Ramakrishnan, 1996). Given its high percentage of forest land i.e. 72.86% compared to 21.34% in the country as a whole, the area has a major advantage in forest based activities and

Table 3.2 Geographical and cultivable area in the Northeastern states during 2015–16

State	Total geographical area (000' ha)	Total cropped area (000' ha)	Net sown area (000' ha)	Cultivable land/arable land/ agri. land		Cropping intensity
				Area (000' ha)	Percentage of geographical area	
1	2	5	7	8	9	10
Arunachal Pradesh	8374	301	227	423	5.05	132.9
Assam	7844	4060	2801	3343	42.62	144.9
Manipur	2233	437	437	444	19.88	100.0
Meghalaya	2243	303	245	1008	44.94	123.3
Mizoram	2108	188	145	367	17.41	130.2
Nagaland	1658	504	384	695	41.92	131.2
Sikkim	710	137	77	97	13.66	144.9
Tripura	1049	486	255	272	25.93	190.1
NER total	26,219	6416	4571	6649	26.43	137.2
All India	328,726	197,054	139,506	181,603	55.24	141.3

Source Govt. of India Census, 2011 and North Eastern Council, 2015

biodiversity, according to a Forest Survey of India report, 2015. Forest ownership trends vary considerably across the region. Since ownership in the hilly area rests with local communities, the proportion of reserved forest in those hilly states is very high (Gadgil & Guha, 1992). The Brahmaputra and the Barak valley in Assam is the largest valley land where settled agriculture is confined.

As a result, production dependent on Common Property Resources (CPR) play important role in people's livelihood in all the states of the region. Forest is an important part of CPR. In quantifying CPR generally land 'other than current fallow', 'cultivable waste', 'pastures', and 'protected' and 'unclassified forests' have been used (Chopra et al., 1990). Generally land in the tribal areas categorised into three types, viz., (i) land owned by the village collectively, (ii) land owned by the chief (locally known as *Gaobura*) who distributes it among individual families and (iii) land owned by the individual families. The first two categories are CPRs and the third is private land (Ganguly, 1978). According to all the tribal Customary Laws community land is collective property and the residents of the village are users of the land. According to Chopra et al. (1990), the CPR in the Northeastern region accounted for 44.07% of geographical area, compared to 20.78% in the country as a whole, with large differences among the states. The highest is in Meghalaya where 71.31% of the total geographical area falls under the CPRs while Assam has the lowest at 15.16%. A total of 58.76% and 56.68% respectively of the land in Nagaland and Mizoram is CPRs while 50.74% of Arunachal Pradesh falls under this category. During the period 1998–99 to 2015–16, the percentage of CPR share to geographical area in the region steadily increased, despite the country's overall

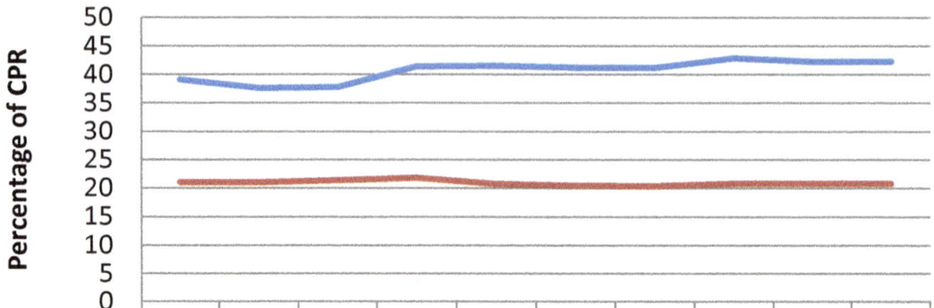

	1998-99	2000-01	2002-03	2004-05	2006-07	2008-09	2010-11	2012-13	2014-15	2015-16
NER	39.15	37.67	37.81	41.42	41.56	41.24	41.28	42.93	42.31	42.31
All India	21.01	21.01	21.39	21.84	20.82	20.46	20.43	20.84	20.83	20.85

Chart 3.1 Percentage of CPR share to geographical area during 1998–99 to 2015–16. *Source* (i) Agricultural Statistics at a Glance, Government of India, Ministry of Agriculture and Farmers Welfare, Department of Agriculture Cooperation and Farmers Welfare, Directorate of Economics and Statistics, various years, (ii) State of Forest Report, Forest Survey of India, various years

decline (refer Chart 3.1). This shows the region's long-term dependence on CPR, which is the primary source of agricultural land and is backed up by a significant number of cultivators and agricultural labourers.

The land holding pattern of the scheduled tribe population of the region is presented in Table 3.3. The figures in the table reveals that the distribution of operating holdings in the area is extremely distorted, with 67.9% of holdings falling into the marginal and limited holding size groups. Despite the fact that seed and fertiliser technology is theoretically scale neutral, it is well recognised that adopting modern agricultural practises may not be financially feasible for small holding sizes (Feder and O' Mara, 1981). The scale-neutrality has been questioned very recently in the case of African small holders (Fischer, 2016). As a result, a sizable portion of the farmers in the area may presently face challenges in implementing and successfully applying new agricultural techniques. The medium and large holding size in the region constituted 14.5% of the total holdings as against 7.2% for the country as a whole. This may be due to distribution of land by the village headmen to the families on the basis of some ideological norm, say family size. Among all the Northeastern states, Assam occupies the highest position both in terms of available cultivable area and net sown area. Approximately 80% of the farmers in the area fall into the marginal (0.40 ha) and small (1.44 ha) categories. Furthermore, as the population grows, the average amount of assets decreases with time. The region's average land holding size (1.60 ha) is slightly more than the average for all of India (1.57 ha). The average land holding size in the region is 0.97 ha in Tripura and 6.92 ha in Nagaland, out of the states. Approximately 55.6% of the holding is just 0.43 ha or smaller. The main reason for this is that large-scale holdings are impractical due to the region's almost two-thirds steep terrain. The average plot size is too small for

Table 3.3 Percentage distribution of number of operational holdings and area operated by scheduled tribes population in Northeastern region during 2010

States	Marginal (0–1 ha)		Small (1–2 ha)		Semi medium (2–4 ha)		Medium (4–10 ha)		Large (10 ha and above)	
	Number	Area	Number	Area	Number	Area	Number	Area	Number	Area
1	2	3	4	5	6	7	8	9	10	11
Arunachal Pradesh	17.9	2.9	17.9	6.8	32.1	24.7	26.2	40.5	5.6	25.3
Assam	59.4	22.4	22.4	26.3	14.4	33.1	3.9	17.0	0.0	1.2
Manipur	45.3	20.3	34.4	34.2	18.8	38.0	1.6	7.6	0.0	0.0
Meghalaya	49.3	15.3	26.0	25.5	20.5	41.8	4.1	16.3	0.0	1.0
Mizoram	54.3	28.6	32.6	36.2	10.9	22.9	2.2	8.6	0.0	3.8
Nagaland	3.4	0.3	11.3	2.1	27.1	11.6	44.1	44.8	14.1	41.2
Sikkim	52.6	12.5	18.4	16.1	15.8	23.2	10.5	33.9	2.6	14.3
Tripura	73.8	33.9	16.5	29.8	8.5	28.2	1.2	7.3	0.0	0.8
NER	46.9	9.9	21.0	13.2	17.5	21.9	11.7	32.1	2.8	22.9
All India	53.9	17.3	24.0	22.6	14.9	26.5	6.3	23.9	0.9	9.7

Source Agricultural Census 2010, Ministry of Agriculture, Government of India.

mechanisation of agriculture and adoption of modern farming practices. Subsistence farming, therefore, is predominant in this region and there is hardly any commercial surplus. Thus according to data available from the Agricultural Census 2010–11, the distribution of operational holdings in the area is highly distorted.

3.5 Forest Coverage

According to the Forest Survey of India report of 2019, the region has a major advantage in forest-based activities and biodiversity due to its high percentage of forest land (68.63%) compared to 21.67% for the country as a whole (refer Table 3.4). Reserved forest, Protected forest, and Unclassified forest are the three classifications for the forest area in general. Although all activities are prohibited unless otherwise permitted in reserved forests, all activities are permitted unless otherwise prohibited in protected forests. Unclassified forest areas, on the other hand, are documented as forest but are not included in reserved or protected areas. The Indian Forest Act, 1927 defined the concept of reserved forest and protected forest. The Forest Survey of India under the Ministry of Environment and Forest has been entrusted with the responsibility of assessment of forest resources in the country. The state of Forest Report published biannually contains Reserved, Protected and Unclassified forests in all the states in the country. The detailed area of three categories of forest is reported in Table 3.5. The data as reported in the table revealed that 54.6% of total reported forest in NER belongs to the category of unclassified forest followed by 34.5% reserved forest and 10.9% protected forest. On the other hand, for India as a whole 56.7% of reported area of forest belongs to reserved forest followed by 28.5% protected forest and 14.8% unclassified forest. The customary laws governs forest resources in many states in the Northeastern region.

The forest biome in the plains varies from tropical to sub-tropical, temperate, and alpine forest habitats in the hills. While the government's conservation and social forestry initiatives, both at the national and state levels, have resulted in an increase in forest area in the country since 1999, they have had no major effect in the region (refer Chart 3.2). According to various Forest Survey of India assessments, the region has lost forest cover to a great extent. According to the latest Forest Survey of India report, the region has lost 765 km^2 of forest cover in 2019 compared to the 2017 assessment (refer Table 3.4 and Chart 3.3). Chart 3.3 also depicted loss and gain account of forest coverage in NER and all India since 1991–2019 in different assessments of FSI, viz., 1991–93, 1997–99, 2003–05, 2011–13, 2017–19 for which published data is available. For the NER in most of the assessments negative change has been reflected implying loss of forest cover in respect of previous assessment.

The per capita availability of forest cover has declined during the period 1997–2019 except 2015–2017 in NER as per the data available from the Forest Survey of India. For the country as a whole it remained stagnant during these periods. However,

Table 3.4 Northeastern states' land cover in 2019

State	Geographical area (GA) (in km^2)	2019 assessment[a] (in km^2)				Forest cover (percentage of GA) 2019	Change in forest cover w.r.t. ISFR 2017	Percentage change in 2019 w.r.t. 2017
		VDF	MDF	OF	Total			
1	2	3	4	5	6	8	9	10
Arunachal Pradesh	83,743	21,095	30,557	15,036	66,688	79.63	– 276	– 0.41
Assam	78,438	2795	10,279	15,253	28,327	36.1	222	0.79
Manipur	22,327	905	6386	9556	16,847	75.46	– 499	– 2.88
Meghalaya	22,429	489	9267	7363	17,119	76.33	– 27	– 0.16
Mizoram	21,081	157	5801	12,048	18,006	85.41	– 180	– 0.99
Nagaland	16,579	1273	4534	6679	12,486	75.31	– 3	– 0.02
Sikkim	7096	1102	1552	688	3342	47.10	– 2	– 0.06
Tripura	10,486	654	5236	1836	7726	73.68	00	00
Northeast India total	262,179	28,470	73,612	68,459	170,541	68.63	– 765	– 0.47
All India	3,287,469	99,278	308,472	304,499	712,249	21.67	3976	0.56

Source State of Forest Report, Forest Survey of India, 2019

[a] *VDF* Very Dense Forest; *MDF* Moderate Dense Forest; *OF* Open Forest

Table 3.5 Recorded forest areas (RFA) in Northeastern states and all India, 2019 (area in km^2)

S. No.	State	Geographical area (GA)	RFA in different categories				RFA as % of GA
			RF	PF	UF	Total RFA	
1	2	3	4	5	6	7	8
1.	Arunachal Pradesh	83,743	10,589	9779	31,039	51,407	61.39
2.	Assam	78,438	17,864	0	8968	26,832	34.21
3.	Manipur	22,327	1467	4171	11,780	17,418	78.01
4.	Meghalaya	22,429	1113	12	8371	9496	42.34
5.	Mizoram	21,081	4483	0	1158	5641	26.76
6.	Nagaland	16,579	234	0	8389	8623	52.01
7.	Sikkim	7096	5452	389	0	5841	82.31
8.	Tripura	10,486	4175	2	2117	6294	60.02
NER total		262,179	34.5[a]	10.9[a]	54.6[a]	131,552	54.63
All India		3,287,469	56.7[a]	28.5[a]	14.8[a]	767,419	23.34

Source India State Forest Report, 2019, Govt. of India
RF reserved forest; *PF* protected forest; *UF* unclassified forest
[a] Percentage of recorded forest area in NER and all India respectively

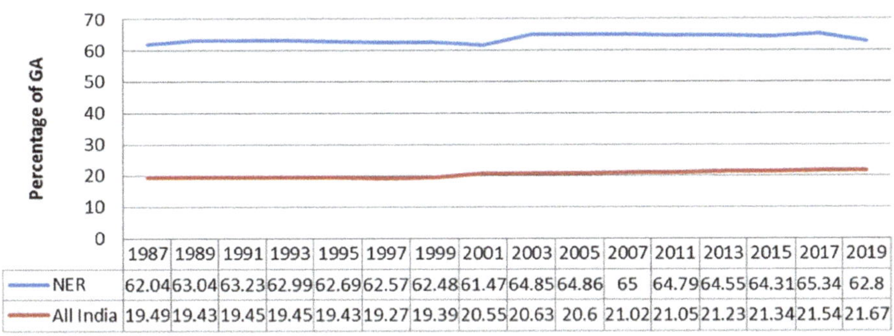

Chart 3.2 Forest cover as percentage of geographical area (1987–2019). *Source* State of Forest Report, Forest Survey of India, 1987–2019 (biannual reports)

for the Northeastern states the per-capita availability of forest has remained higher as compared to all India level over the period under study (for details refer Chart 3.4).

Another important aspect of climate change is tree cover, which consisted of tree patches outside the recorded forest area exclusive of forest cover and less than the minimum mappable area (1 ha) as survey conducted by Forest Survey of India. Interestingly the tree cover as percentage to geographical area in the NER has remain less than that of all India average during the period 2000 to 2017–18. This is to be noted that Northeastern states occupy highest percentages of forest cover in the country as a whole. Perhaps due to this highest forest cover the percentage of tree

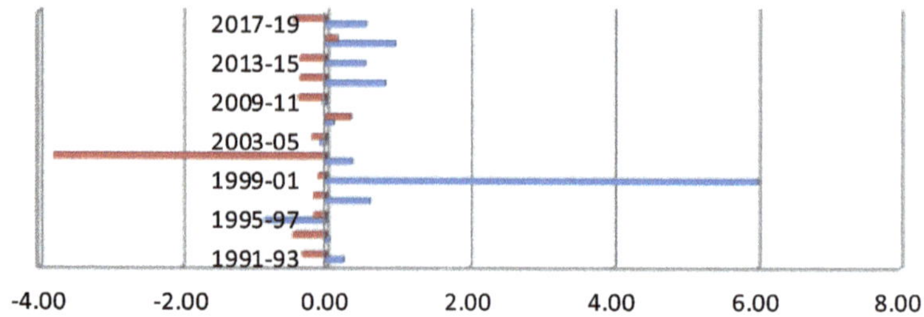

	1991-93	1993-95	1995-97	1997-99	1999-01	2001-03	2003-05	2005-07	2009-11	2011-13	2013-15	2015-17	2017-19
NER	-0.33	-0.46	-0.16	-0.17	-0.11	-3.80	-0.19	0.35	-0.38	-0.36	-0.36	0.16	-0.45
All India	0.24	0.04	-0.86	0.62	6.00	0.36	-0.09	0.11	-0.05	0.83	0.54	0.97	0.56

Percentage of loss/gain

Chart 3.3 Trends in loss/gain in of forest cover in assessments of forest survey (in percentage of previous assessment of respective period) during 1991–1993 to 2017–2019. *Source* State of Forest Report, Forest Survey of India, 1987–2019 (biannual reports)

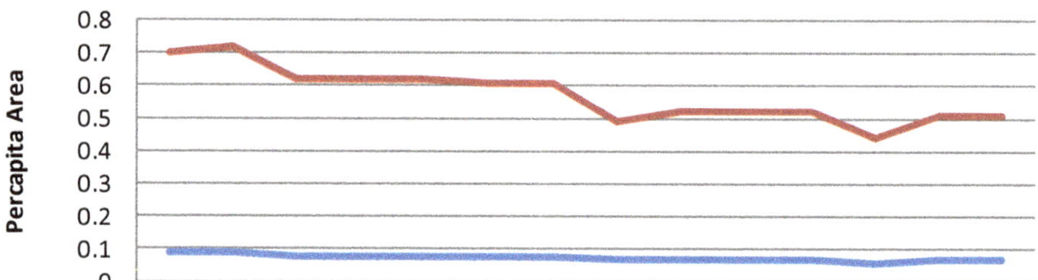

	1987	1989	1991	1993	1995	1997	1999	2001	2003	2005	2007	2011	2013	2015
NER	0.61	0.63	0.54	0.54	0.54	0.53	0.53	0.42	0.45	0.45	0.45	0.38	0.44	0.44
India	0.09	0.09	0.08	0.08	0.08	0.08	0.08	0.07	0.07	0.07	0.07	0.06	0.07	0.07

Chart 3.4 Per-capita availability of forest (1987–2015) (in hectare). *Source* State of Forest Report, Forest Survey of India, 1987–2015 (biannual reports)

cover has been less in the region. The detailed account on tree cover as percentage of geographicial area for the Northeastern states is depicted in Table 3.6 and Chart 3.5.

3.6 Governance of Forest Resources

Forest conflicts have a long history in the hill regions of Northeastern India, dating back to inter-tribal conflicts that have existed regularly since the area was settled over 1000 years ago. Tribal opposition to British colonial incursions into Northeast India's

Table 3.6 State-wise tree cover as percentage of geographical area

State	2000	2002	2004	2006	2008–09	2010–11	2013–14	2015–16	2017–18
1	2	3	4	5	6	7	8	9	10
Arunachal Pradesh	0.57	0.43	0.53	0.71	0.66	0.79	0.91	0.96	1.01
Assam	2.48	1.19	1.89	2.03	1.99	2.02	2.06	1.91	1.80
Manipur	0.43	0.61	0.64	0.88	0.86	1.00	1.09	0.99	0.77
Meghalaya	0.62	1.57	1.81	2.42	2.58	2.98	3.17	2.93	3.17
Mizoram	0.45	0.62	0.58	0.82	0.90	1.06	2.54	2.22	2.09
Nagaland	0.42	1.31	1.44	1.81	1.94	2.24	2.30	2.29	2.18
Sikkim	0.20	0.31	0.38	0.28	0.35	0.44	0.49	0.49	0.51
Tripura	0.65	1.11	1.28	1.63	1.75	2.03	2.22	2.05	2.20
NER total	1.11	0.87	1.14	1.37	1.38	1.52	1.72	1.63	1.61
All India	2.48	3.04	2.79	2.82	2.76	2.78	2.82	2.85	2.89

Source India State of Forest Report—2001–2019, Forest Survey of India, Ministry of Environment, Forest and Climate Change

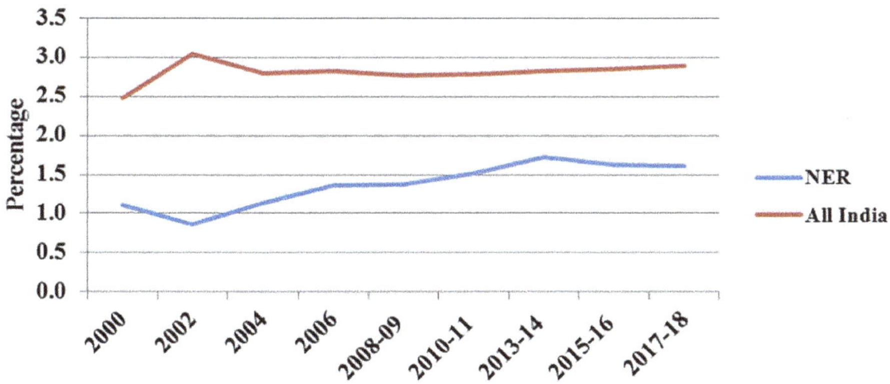

Chart 3.5 Tree cover as percentage of geographical area

hills resulted in special policies enabling customary forest management systems and reverence for traditional governance systems in the early nineteenth century. This policy reflected the British colonial government's acceptance that the hill communities could not be managed centrally and were better off functioning under their own governance systems. Furthermore, the Indian constitution acknowledges the unique rights of most indigenous hill groups under the Sixth Amendment, 1956. The Government of India has attempted to establish an integrated set of national environmental and development policies since independence, especially in relation to the management of public forestlands, but the states of the Northeast have specific needs and problems. Perhaps no other area exemplifies this more than community based forest management.

Unlike most of the Indian subcontinent, where forest departments have acted as state forest managers for over a century, most state forest departments in Northeast India have only existed since the 1970s. At least two-thirds of the area's forests are legally under the control of autonomous district councils and are physically owned and controlled by rural citizens. Village councils, chieftainships, and elderly councils, which are focused on a tiny, homogeneous village culture that supports shared needs and interests, have historically done a good job of maintaining their forest resources (Ostrom, 1990). Unfortunately, the historical background and recent experiences of indigenous cultural institutions in forest management have been under reported, and institutions have earned very little external funding from state or national governments or international agencies. Based on the legislation passed during the British colonial period, as well as the Indian Constitution and laws approved in the 1950s in particular, the Northeast's forest-dependent communities have considerable legal authority over the region's natural resources. However, these resources are currently threatened by the absence of systematic mapping, boundary registration and titling of indigenous community based forest management systems. Trends in privatisation, agricultural sedentarisation and commercialisation have all guided deforestation promoting practices (Dasgupta, 2005). In some areas the community values have largely weakened, supporting forest protection and sustainable use. Singh (1996) observed that "this lack of understanding is ... visible in national policies and programs that ignore exceptional circumstances, resulting in a waste of resources to the disadvantage of the very people they are intended to benefit".

3.7 Cropping Pattern and Agricultural Growth

A striking feature of the cropping pattern of the Northeastern region is the predominance of cereals, mainly paddy, which accounts for 59.28% of total cropped area in 2015–16 as against 83.25% in 1990–91 (refer Table 3.7). Thus the cropping pattern in the region undergoes significant changes since 1990–91. However, total food crops constituted 83.65% in the NER as against 71.56% for the country as a whole in 2015–16. This indicates slow process of diversification of agricultural sector in the region. Another important reason for slow rate of diversification is the non-availability of irrigated area in the NER. While the net irrigated area in the Northeastern region steadily decined from 24.36% in 1950–51 to 10.37% in 2012–13, the same increased from 15.43 to 47.67% for the country as a whole during the same period.

For the country as a whole significant breakthrough in agricultural sector occured during mid 1960s with the use of seed-water-fertiliser technology. The country became food surplus economy from a food deficit one by the beginning of 1970s. The share of pulses and other non-food crops shows an increasing trend for the country as a whole since 1970–71. But opposite trend observed for the Northeastern states of

Table 3.7 Percentage share of area under different crops in total cropped area during 1950–51 to 2015–16

Year	Region/India	Cereals	Pulses	Other food crops[a]	Total food crops	Non-food crops
1	2	3	4	5	6	7
1950–51	NER	85.43	2.65	0	88.08	11.92
	All India	75.32	12.41	1.1	88.83	11.17
1970–71	NER	85.25	2.87	1.12	89.24	10.76
	All India	74.53	11.37	1.32	87.22	12.78
1990–91	NER	83.25	3.63	1.02	87.90	12.10
	All India	69.59	12.30	1.84	83.73	16.27
2015–16	NER	59.01	4.02	19.59	82.62	17.38
	All India	48.82	11.32	12.64	72.78	27.22

Source Department of Agriculture, Cooperation and Farmers Welfare, Directorate of Economics and Statistics, Govt. of India

[a] Other food crops include sugarcane, fruits, spices and vegetables

India. Even during the period 2005–16 to 2014–15 there has been a dismal scenario of availability of irrigation facilities in the north eastern region in constrast to the entire country (refer Charts 3.6 and 3.7).

The annual compound growth rates of area, production, yield and CPR estimated for the period 2000–01 to 2013–14 are presented in Table 3.8. The area growth for both the foodgrains and non-foodgrains mainly attributed to growth of CPR area in NER and growth of yield estimated only 1.2%. For the country as a whole, negative growth of food grains area was compensated by higher growth of area under non-foodgrains (2.2%) and yield growth (3.4%).

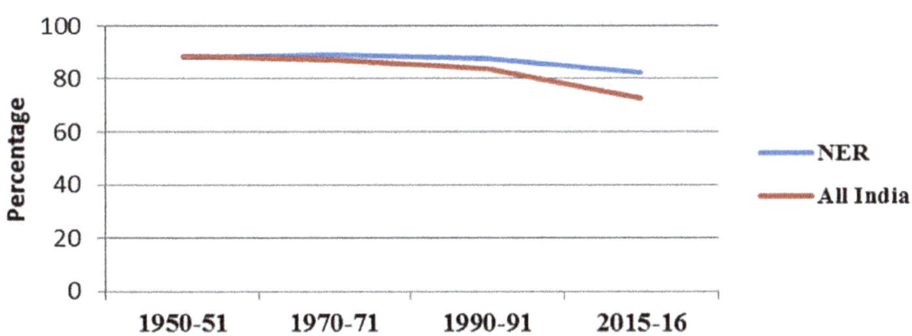

Chart 3.6 Trend of area under food crops (in percentage). *Source* Department of Agriculture, Cooperation and Farmers Welfare, Directorate of Economics and Statistics, Govt. of India

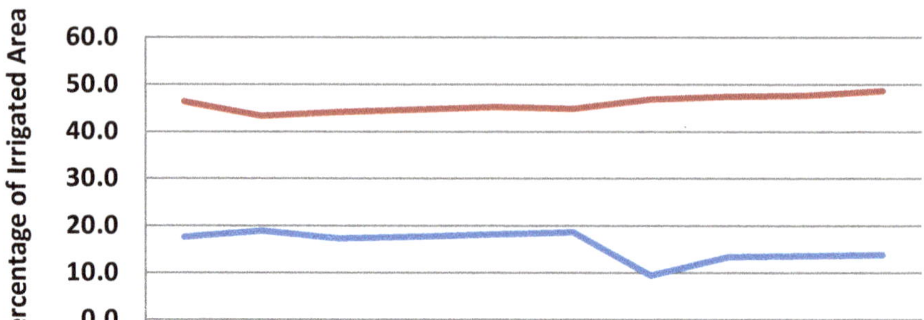

	2005-06	2006-07	2007-08	2008-09	2009-10	2010-11	2011-12	2012-13	2013-14	2014-15
NER	17.7	19.0	17.4	17.7	18.2	18.6	9.5	13.4	13.6	13.8
All India	46.4	43.4	44.2	44.7	45.2	44.9	46.9	47.5	47.7	48.6

Chart 3.7 Percentage of net irrigated area to net cultivated area 2005–06 to 2014–15. *Source* (i) Directorate of Economics and Statistics, Ministry of Agriculture and Farmers Welfare, Govt. of India, (ii) The Fertiliser Association of India, accessed at Indiastat.com

3.8 Use of Soil Nutrients

Data on consumption of chemical fertilizers combining Nitrogen (N), Phosphorus (P), and Potassium (K) revealed that fertilizer consumption per hectare was higher in Tripura (46.11 kg) followed by 41.26 kg in Assam, 34.72 kg in Nagaland, 34.68 kg in Manipur during 2016–17, while in Arunachal Pradesh it was abysmally low. It follows that fertilizer use had almost a complementary relationship with the availability of irrigation facilities. There is no chemical fertilizer consumption in the state of Sikkim. The shifting cultivation practices is also absent in the state of Sikkim. All the Northeastern states had less fertilizer consumption than all India level during the period 2012–13 to 2016–17 (refer Table 3.9 and Chart 3.8).

As evident from the analysis of data presented in Table 3.9 chemical fertilizer (N + P + K) consumption in the Northeastern states has been dismally low compared to all India level. But one of the encouraging developments in respect of fertilizer consumption in the region has been the significant increase in the Bio-fertiliser. Data from Table 3.10 indicated that the production of bio-fertiliser (Carrier based) has increased from 0.44% as compared to all India production in 2014–15 to 2.53% in 2017–18 though declined to 0.84% in 2018–19. Similarly in the year 2014–15 the NER did not produce any liquid based bio-fertiliser. The production of liquid based bio-fertilizer was started in the states of Assam and Sikkim in 2015–16 and consisted 3.61% of all India production. Significantly in 2018–19 the liquid based bio-fertiliser production consisted of 87.92% of all India production, mainly concentrated in the state of Sikkim. In fact, Sikkim has completely switched over to completely organic fertiliser consumption and played a very important role in all India level in this respect.

Table 3.8 Annual compound growth rate during 2000–01 to 2013–14

States		Area	Production	Yield[a]	CPR
1	2	3	4	5	6
Percentage					
Arunachal Pradesh	Foodgrains	1.2	4.7	4.0	2.0
	Non-foodgrains	2.4	5.3		
Assam	Foodgrains	– 0.2	2.2	2.6	– 0.1
	Non-foodgrains	0.7	0.6		
Manipur	Foodgrains	3.9	1.9	– 1.0	0
	Non-foodgrains	32.2	27.3		
Meghalaya	Foodgrains	0.0	2.0	2.1	– 0.5
	Non-foodgrains	1.1	0.5		
Mizoram	Foodgrains	– 4.9	– 7.8	– 3.6	0.1
	Non-foodgrains	– 7.9	– 2.5		
Nagaland	Foodgrains	2.0	4.4	2.6	1.1
	Non-foodgrains	3.2	7.1		
Sikkim	Foodgrains	– 0.7	0.6	1.5	– 5.0
	Non-foodgrains	– 1.1	0.7		
Tripura	Foodgrains	0.3	2.5	2.1	– 2.1
	Non-foodgrains	4.1	– 3.0		
NER	Foodgrains	0.2	2.3	1.2	0.7
	Non-foodgrains	1.7	2.2		
All India	Foodgrains	– 1.4	2.5	3.4	– 0.3
	Non-foodgrains	2.2	2.7		

Source Directorate of Economics and Statistics, Ministry of Agriculture and Farmers Welfare, Govt. of India

[a] Only foodgrains considered

In concomitant with the higher production of bio-fertilizer, mainly liquid based in the Northeastern region the production of organic manure has also been recorded significant proportion in comparison with the country as a whole. During the year 2017–18 the production of organic manure consisted of 7.44% of all India total (for details refer Table 3.11).

The area under organic certification process combining cultivated and wild harvest in the Northeastern region has undergone significant changes from 2.3% of all India figure in 2014–15 to 5.5% in 2018–19.

Among the Northeastern states Sikkim occupy a very prominent position in respect of area under organic certification consisting 40% of total area of the region (refer Table 3.12).

The area under the organic farming in the Northeastern states consisted of 5.5% of total area for the country as a whole in the year 2018–19. The figures for total area

Table 3.9 State-wise consumption of fertilizers (N + P + K) (kg/ha) during 2012–13 to 2016–17

State	2012–13	2013–14	2014–15	2015–16	2016–17
1	2	3	4	5	6
Arunachal Pradesh	2.01	0	0	2.06	0
Assam	65.67	88.78	125.08	44.79	41.26
Manipur	35.08	44.21	61.84	41.62	34.68
Meghalaya	14.26	47.26	0		
Mizoram	15.18	–	–	18.18	22.88
Nagaland	4.44	5.17	6.34	5.96	34.72
Sikkim	0	–	–	–	0
Tripura	69.13	50.65	41.15	43.00	46.11
NER average	33.96	47.21	58.6	30.71	35.93
All India average	131.37	118.55	128.08	130.66	123.41

Source Agricultural Statistics at Glance 2018, Department of Agriculture Cooperation and Farmers Welfare, Ministry of Agriculture and Farmers Welfare, Govt. of India

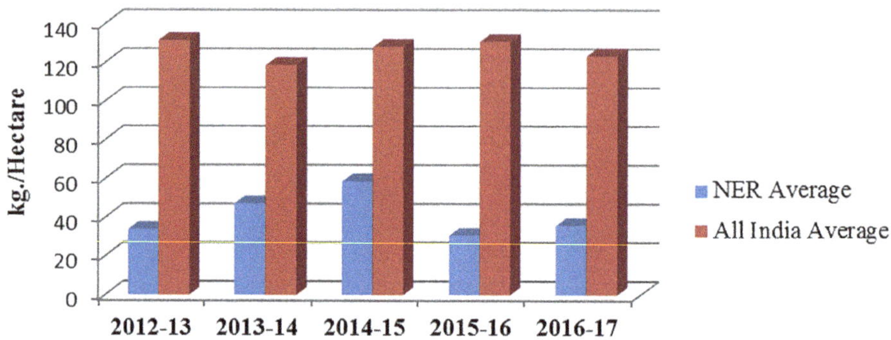

Chart 3.8 Consumption of fertilizers (N + P + K) (kg/ha). *Source* Agricultural Statistics at Glance 2018, Department of Agriculture Cooperation and Farmers Welfare, Ministry of Agriculture and Farmers Welfare, Govt. of India

of organic farming as depicted in Table 3.13 which combines four categories of area, viz., (i) Certified cultivated organic Area, (ii) In-conversion cultivated area, (iii) Total area cultivated under certification process and (iv) Wild area. Though total area under organic farming in NER was 5.5% of all India total, the area under in-conversion of cultivated area consisted of 10.8% followed by 9.5% of total area cultivated under certification process, 8.6% of certified cultivated organic area and 0.1% of wild area of all India figures. There is higher concentration of area under organic farming in the state of Sikkim to the tune of 40% of total area of NER.

Table 3.10 State-wise production of bio-fertilizers during 2014–15 to 2018–19

State	Carrier based (in metric tonnes)					Liquid based (in KL)				
	2014–15	2015–16	2016–17	2017–18	2018–19	2014–15	2015–16	2016–17	2017–18	2018–19
1	2	3	4	5		6	7	8	9	10
Arunachal Pradesh	59.0	1180	119.70	232.90	0.0	0.0	0.0	0.0	0.11	0.0
Assam	88.0	1315	1359.05	1742.6	617.20	0.0	225	26.0	33.10	–
Manipur	0.0	0.0	25.0	23.94	18.0	0.0	0.0	0.0	–	4.0
Meghalaya	0.0	0.0	0.0	–	–	0.0	0.0	0.0	–	–
Mizoram	3.6	4.2	2.5	8.99	2.50	0.0	0.0	0.0	–	–
Nagaland	7.45	8.81	51.45	70.62	17.45	0.0	0.0	0.0	–	–
Sikkim	12.4	12.91	16.25	33.15	–	0.0	0.0	0.0	–	51,500
Tripura	240.0	1143.07	1153.50	1187.88	81.79	0.0	0.0	0.0	0.03	–
NER average	351.45 (0.44)	2483.99 (2.82)	2607.75 (2.39)	3067.18 (2.53)	736.94 (0.84)	0 (0.0)	225 (3.61)	26 (0.35)	33.13 (0.37)	51,504 (87.92)
All India average	80,696.45	88,029.30	109,020.1	121,066.5	87,919.5	4054.56	6240.92	7526.3	9033.1	58,578.0

Source Annual Report 2018–19, National Center of Organic Farming, Department of Agriculture Cooperation and Farmers Welfare
Figure in the parentheses represent percentage of all India consumption

Table 3.11 State wise production of organic manures, 2017–18

State	Organic manure	Farm yard manure	City compost	Vermi-compost	Phosphate rich organic manure (PROM)	Others	Total manure lakh MT
1	2	3	4	5	6	7	8
Arunachal Pradesh	1.6	0.35	–	1.89	0.04	1.56	5.44
Assam	–	98.4	–	98.6	0.01	3.24	200.25
Manipur	–	0.98	–	0.09	–	–	1.07
Meghalaya	–	1.98	–	1.13	–	–	3.11
Mizoram	–	23.1	–	9.7	–	–	32.8
Nagaland	0.64	2.2	–	1.23	–	–	4.07
Sikkim	0.45	–	–	0.09	0.51	0.04	1.09
Tripura	4.12	–	–	0.13	–	–	4.25
NER average	6.81 (1.80)	127.01 (6.73)	0 (0.0)	112.86 (14.56)	0.56 (9.49)	4.84 (2.08)	252.08 (7.44)
All India average	377.36	1885.92	110.46	775.01	5.90	232.55	3387.20

Source Annual Report 2016–17, National Center of Organic Farming, Department of Agriculture Cooperation and Farmers Welfare

Table 3.12 Total area under organic certification process during last 5 years (cultivated + wild harvest) (area in hectares) during 2014–15 to 2018–19

State	2014–15	2015–16	2016–17	2017–18	2018–19
1	2	3	4	5	6
Arunachal Pradesh	3688.61	72,485.26	72,311.27	6179.69	9246.94
Assam	16,258.02	28,493.24	23,930.40	28,071.81	28,234.67
Manipur	168.2	251.4	241.4	5397.9	7460.82
Meghalaya	4489.29	4609.42	9629.60	40,335.66	48,409.74
Mizoram	764.24	213.8	210	998.95	7039.89
Nagaland	8362.43	6186.93	4699.93	8839.86	8268.56
Sikkim	76,392.38	75,851.21	75,218.28	76,076.18	75,798.92
Tripura	203.56	203.56	203.56	2251.19	2534.52
NER total	110,326.7 (2.3)	188,294.8 (3.3)	186,444.4 (4.2)	168,151.2 (4.7)	186,994.1 (5.5)
All India	4,893,851	5,710,384	4,452,987	3,566,538	3,428,638.77

Source http://apeda.gov.in (Agriculture and Processed Food Products Export Development Authority, Ministry of Commerce and Industry, Government of India as compiled from EnviStats India 2020, Vol. I, Social Statistics Division, Ministry of Statistics and Programme Implementation, Govt. of India)

Figure in the parentheses refer to percentage of all India

Table 3.13 State-wise area under organic farming (registered under accredited certification bodies) (area in hectares), 2018–19

State	Certified cultivated organic area	In conversion cultivated area	Total area cultivated under certification process	Wild area	Total cultivated + wild
1	2	3	4	5	6
Arunachal Pradesh	627.2	8619.8	9246.9	–	9246.9
Assam	15,223.5	12,951.2	28,174.7	60	28,234.7
Manipur	241.4	5219.4	5460.8	2000	7460.8
Meghalaya	1612.7	46,797.0	48,409.7	–	48,409.7
Mizoram	0.0	7039.9	7039.9	–	7039.9
Nagaland	2751.2	5517.4	8268.6	–	8268.6
Sikkim	73,654.9	2144.0	75,798.9	–	75,798.9
Tripura	203.6	2331.0	2534.5	–	2534.5
NER total	94,314.5 (8.6)	90,619.7 (10.8)	184,934 (9.5)	2060 (0.1)	186,994 (5.5)
All India	1,097,074.4	841,146.4	1,938,220.8	1,490,417.98	3,428,638.8

Source Agriculture and Processed Food Products Export Development Authority, Ministry of 'Commerce and Industry, Government of India; http://apeda.gov.in as compiled from EnviStats India 2020, Vol. I, Social Statistics Division, Ministry of Statistics and Programme Implementation, Govt. of India
Figure in the parentheses refer to percentage of all India

3.9 Livelihood Pattern of Northeast India

In the Northeastern region, the pace of development differs significantly between plains and hilly regions. The valleys of the area are economically active, with the Brahmaputra valley being the most active. The tribals are the majority in four states, viz., Mizoram, Meghalaya, Nagaland, and Arunachal Pradesh. In these states tribal population make up as much as 95% of the population. Though the region is diverse and heterogeneous, it is also homogeneous in the sense that it lacks the social stratification that occurs in other parts of the world.

In terms of resource endowments, extent of industrialisation, and infrastructure facilities, the states of the Northeastern region vary. Tea and timber have dominated the industrial sector in Assam, while mining, saw mills and plywood factories have flourished in other parts of the region. Although it has not yet reached its full potential, the region's economy is still primarily based on agriculture. The service sector is facing a shortage of jobs because industry and agriculture have not really taken off despite the region's large untapped resource base and capacity.

With the exception of Meghalaya and Nagaland, this region's state domestic income is largely derived on agriculture. The NSDP is heavily influenced by forestry and logging in Arunachal Pradesh and Nagaland, as well as mining in Meghalaya. Construction also makes a major contribution in this field. Since the 1990s, Assam's per capita income growth has been almost stagnant, while Meghalaya has out performed the national average, Arunachal Pradesh is on the rise, and Manipur, Nagaland, and Tripura have made little development.

As compared to the national average, the population growth rate in the Northeastern states is much higher. Some states, such as Assam, Tripura, and to a lesser extent, Meghalaya and Mizoram, have historically seen large-scale influxes of people from neighbouring countries such as Bangladesh and Nepal, while others have a much higher birth rate than death rate. As a result, the Northeastern states' decadal change in the population tends to be much higher than all India average.

The region's population has a higher overall literacy rate than the national average. Mizoram has the country's second highest literacy rate. Another important demographic characteristic of the region is the country's higher concentration of scheduled tribes. It ranges between 85% and 95% in the states of Arunachal Pradesh, Meghalaya, Mizoram, and Nagaland. Cutivators and farm workers made up more than half of the working population, comparable to the rest of India (refer Table 3.9). Though the region's working population and key worker ratios are higher than the national average, cultivators account for 35.49% of the region's population, compared to 24.63% in India. Agricultural labourers, on the other hand, make up 14.74% of the population in the area, compared to 29.95% in India overall, according to the 2011 census. As a result, the region's economy is agrarian, and agriculture is the primary source of income for the majority of its inhabitants (refer Table 3.14).

The administrative structure has serious implications on the land relations (Hansaria, 2005). In this administrative system, the definition of regional common property resources is crucial. The tribal tradition conferred the right of ownership of the CPR mostly on the communities. The traditional rules and regulations have not been adapted to the formal laws of transfer of ownership.

Due to the prevailing topography, the farmers in the region are dependent mostly on jhum cultivation. Common Property Resources (CPR) play an important role in people's livelihood in the Northeastern region. Forest is an important part of CPR. The CPR in the Northeastern region constituted 44.07% of geographical area as against 20.78% in the country as a whole with wide variations amongst the states. This indicates livelihood dependency of the region on CPR over the period, which is main source of agricultural land and substantiated by large percentage of cultivators and agricultural labourers.

Table 3.14 Classification of workers in NER and all India, 2001 and 2011 (in percentage)

States	Total worker 2011			Main worker as percentage of total worker, 2011			Cultivators (main and marginal)		Agricultural labourers (main and marginal)	
1	2	3	4	5	6	7	8	9	10	11
	Total	Male	Female	Total	Male	Female	2001	2011	2001	2011
Arunachal Pradesh	42.47	59.61	40.39	81.46	51.24	30.22	57.8	51.51	3.9	6.16
Assam	38.36	71.36	28.64	72.58	58.77	13.81	39.1	33.93	13.2	15.42
Manipur	45.68	56.68	43.32	74.67	47.15	27.52	40.2	35.10	12.0	8.51
Meghalaya	39.96	59.35	40.65	77.73	49.39	28.34	48.1	41.72	17.7	16.73
Nagaland	49.24	56.19	43.81	76.09	45.40	30.69	64.7	55.20	3.6	6.46
Mizoram	44.36	59.74	40.26	85.27	54.10	31.17	54.9	47.17	5.7	8.59
Sikkim	50.47	63.07	36.93	74.77	69.67	30.33	28.11	24.91	4.27	6.62
Tripura	40.00	71.13	28.87	73.29	60.42	12.87	27.0	20.14	23.8	24.06
Northeast India	42.87	62.01	37.99	77.30	52.35	24.95	47.4	35.49	11.41	14.74
India	39.80	68.88	31.12	75.24	56.70	18.54	31.7	24.63	26.5	29.95

Source Registrar General of India 2011, Agricultural Statistics at a Glance, 2014, Ministry of Agriculture, Govt. of India

References

Agarwal, A. K. (1996). Agro industries for sustainable development of tribal areas in North East India. *Political Economy Journal of India, 8*(3–4), 127–131.

Chopra, K., Kadekodi, G. K., & Murthy, M. N. (1990). *Participatory development: People and common property resources.* Sage.

Dasgupta, P. (2005, December 5–16). Common property resources: Economic analytics. In *Teaching Workshop on Environmental Economics for the Middle East and North Africa*, ICTP, Trieste, Italy.

Feder, G., & O'Mara, G. T. (1981). *Farm size and the diffusion of green revolution technology* (World Bank Reprint Series, No. 207).

Fischer, K. (2016). Why new crop technology is not scale-neutral—A critique of the expectations for a crop-based African Green Revolution. *Research Policy, 45*(6), 1185–1194.

Gadgil, M., & Guha, R. (1992). *The fissure land: An ecological history of India.* Oxford University Press.

Ganguly, J. B. (1978). Socio-economic transition of shifting to sedantary cultivation in North East India. In NEICSSR (Ed.), *Shifting cultivation in North East India.* Shillong.

Hansaria, B. L. (2005). *Sixth schedule to the constitution* (2nd Edition revised by Vijay Hansaria). Universal Law Publishing Company. New Delhi.

Medhi, B. K., Athparia, R. P., & Jose, S. V. D. K. (2009). *Tribes of North-East India, issues and challenges.* Omsons Publications.

Ostrom, E. (1990). *Governing the commons: The evolution of institutions* (pp. 2–9). Cambridge University Press.

Ramakrishnan, P. S. (1996). *Shifting agriculture and sustainable development.* UNESCO and The Parthenon Publishing Group.

Singh, S. (1996). Joint forest management in India. In E. Lutz & J. Caldecott (Eds.), *Decentralization and biodiversity conservation.* The World Bank.

Chapter 4
Shifting Cultivation, Land Use Land Cover and Soil Erosion in Northeastern States

4.1 Framework of Shifting Cultivation Practices

Shifting cultivation is believed in originated in the neolithic period around 7000 B.C. in Northeastern hilly region (Borthakur, 1992). It is also alternately called as the slash and burn method of cultivation. It is the natural way of life to the tribal population of the region. In fact, 'jhumming' characterised the bulk of agriculture in the uplands of the Northeastern region. At the commencement of the twentieth century limited population enabled 'jhummias' to move to vergin unjhummed areas, every third or fourth year. The jhum cycle thus covered a 15–20 years period, allowing once jhummed areas to regenerate enough vegetative growth to sustain fertility for the next round of jhum cultivation. Available records reveal that jhum cultivation had, by then, degraded not more than 10% of the vegetation and landmass of the region and yielded productivity of one metric tonne of paddy per hectare. By 1950s, due to multiplication of population the jhum cycle had reduced to a 10–15 years period, the same piece of jhum land could not be cultivated for more than two years at a stretch. By 1951, jhum cultivation had degraded about 15% of the region's vegetation and landmass and yield had reduced to 0.6–0.7 metric tonnes per hectare. The independent Government. In India rightly accorded high priority to a new Jhum Control programme, and a large number of schemes were launched to assist shifting cultivators to sustainable and permanent methods of cultivation. As per the statistics available from the North Eastern Council by the period 2000 about 25% of the landmass and vegetation of the region had degraded with a reduction in the jhum cycle of 4–6 years. The productivity of jhum land has fallen to about 0.5 metric tonnes per hectare.

Thus, though traditional jhumming was the only economic occupation of almost all the tribals of the region, it has lost its past status among the jhummias of the region. The situation of jhumming vis-a-vis the other occupations and the different means of living have changed for the shifting cultivators of the region. Even though jhumming has a different standing in the Northeastern region now that jhummias

N. Roy et al., *Climate Change, Shifting Cultivation and Livelihood Vulnerabilities in India*, https://doi.org/10.1007/978-3-031-54927-4_4

have other sources of income and multiple jobs in addition to jhumming, the region's farmers still rank among the lowest of the poor (Bose, 1991).

4.2 Method of Shifting Cultivation

The forest farmer in the humid tropics practised for centuries the method of 'shifting agriculture' or popularly known in India as 'Jhum' (Ramakrishnan, 1996). This is the simplest method available to the tribal population living in the hilly region. There is no other alternative way of cultivating the land in the hill area. The shifting cultivation is the natural way of life to the tribal people in the hill area. In fact, the man–nature relationship of a community is reflected in its mode of resource use across the ideological axis. At a given level of ideology, man's material requirements and his perception of nature and natural phenomena determines his interaction with nature and characterises the ideal type of the model of resource use. At this level, ideology acts as a determinant variable of the mode of resource use and is external to the system. The ideological base of a community, for example, governs production relations, forces of production and exploitation of natural resources as well (Gadgil & Guha, 1992). Use of land resources for shifting cultivation by the different tribes is, however, not indiscriminate, instead follows certain norms at the ideological level.

The system of shifting cultivation involves a number of operations. First, forest bushes are cut up to the stump level in the months of December–January, and small trees and bamboos are felled. Short tree stumps and large tree boles are left intact. The underground organs of different species are not disturbed. Next, the stumps are allowed to dry, after which they are set on fire. After the burning is completed, the land is cleared and digging stick and seeds of different crops as required are scattered into the holes before the onset of the rains. The cultivation is confined to the area close to the village. After every crop season or two, the area is left for 25 or 30 years, and the cultivator moves to another location to slash and burn and cultivate. This laborious process often completed by the men from two or three families. Such a joint effort is one of the essential ingredients of a well-knit social organisation. This effort along with the process of allotment of sites for jhum by the village headman who is in charge of overall control of the village community (known as Gaobura) helps to promote kinship among the members of village (Ramakrishnan, 1996). However, the system of allotment of lands differs from tribes to tribes.

Paddy is the main crop in jhumming. From the available data it has been observed that about 90% of the total cultivable area in the hills of the Northeastern region belongs to paddy. The present study thus confirmed 'monocropping' nature of cultivation. Among other crops cultivated are maize, millet, sweet potatoes, leafy vegetables etc. Each family chooses its own cropping pattern according to what it consumes and quantities enough to meet its own needs. The crops are all rainfed, and harvesting starts in August. The paddy grown in jhum fields matures earlier than the paddy grown in plains. It has been confirmed by different studies of the farmers practising shifting and settled cultivation.

4.3 Jhum Cycle

Jhum or shifting cultivation had created to a cyclic process popularly known as Jhum Cycle. Under jhum cultivation, a cycle is completed by cultivating in one area for one or two years, then shifting to another area for one or two years, and then returning to the first area for cultivation. A Jhum cycle is, thus, defined as the number of years between two consecutive jhum operations. The interval involves the number of years a field is cultivated as well as the number of years it is left fallow. Previously the fallow period was about 25–30 years but now it has been has reduced the cycle to 4–5 years. Traditionally the jhumias cultivate plot for one or two years or as long as they get sufficient yield and leave the area in search of new plot in another section of forest as and when they feel it necessary. Earlier, when the number of jhumias was less, many of them never returned to a plot second time. However, situation in a major part of the forest is rather critical at present. The pressure of increasing population is felt everywhere. And the number of people who depend on jhum cultivation has also increased visibly. The jhumias can no longer afford a gap of 10 or more years before they return to a previously cultivated plot. The situation has become so alarming that shifting cultivation in the region now causing some sort of tragedy of commons, since the users of common forest do not manage them adequately.

A number of reasons have been encountered behind shifting cultivation being the primitive way of survival of the tribal in the region. A few important ones are:

(i) Shifting cultivation involves the easiest method of preparing land for cultivation.
(ii) Within a very short period of time crops can be produced and harvested.
(iii) Shifting cultivation is also free from the danger of flood, draught etc.
(iv) Under shifting cultivation there is regular supply of water from mountain stream. Long cycles were possible earlier because population pressure was not heavy, and land availability did not limit the cycle.

As per the estimate from the data by the Task Force on Shifting Cultivation, Ministry of Agriculture (1983) Government of India, it has been observed that the average 'jhum cycle' in 70% of the agricultural farms ranges between 5 and 15 years. Jhumming is responsible for large scale soil erosion, siltation of reservoir, incidence of floods in the plains, loss of soil fertility resulting in low production, dying up of natural streams etc. and also effect the flora and fauna to a great extent.

The growth of human populations in the Northeastern region and the increasing strain that additional migrants put on the region's forests are a major factor affecting forest cover. Although urbanisation has increased rapidly over the past few decades at a rate of 15% to 20% per year in many states, the rural population has also grown. Over the twentieth century, the population of most Northeastern states grew eight to tenfold.

4.4 Concerns for Sustainable Shifting Agriculture

Shifting cultivation provides a very basic way of preparing land for cultivation, which is very beneficial to the hill people. By using the slash and burn method, weeds and jungles can be quickly cleared, and yield can be collected in a short amount of time. The most serious negative impact of shifting cultivation, however, is the loss of forest in hilly areas, which results in severe soil erosion due to rainfall. Furthermore, the depletion of 'topsoil' in hill zones, which accounts for 22% of overall land, produces a significant drop in the earth's fertility, posing a serious problem for tribal societies. The major issues associated with jhum cultivation are the drastic shortening of the 'jhum cycle' to 4–5 years during recent years. The leading cause for shortening the cycle is the rapid increase in the tribal population and consequent reduction in the availability of land for cultivation. The apparent effect of shortening of the process can be grouped into two classes—(a) Economic effects and (b) Environmental impacts. However, the indigenous societies in different states in the region are aware of the consequences of shifting cultivation which was evidenced from the decline in the area of shifting cultivation across all the states in the region during the period 2000 and 2015–16 (refer Table 4.1). Despite the decline of shifting cultivation practices across different states, a large percentage of cropped area is still under shifting cultivation practices. As per the data in Table 4.1, it is revealed that in the state of Mizoram 36.78% of total cropped area is under shifting cultivation followed by 19.43% in Nagaland, 16.91% in Arunachal Pradesh, 11.44% in Manipur, 7.85% in Meghalaya, 0.63% in Tripura and 0.13% in Assam. It is to be noted that in Assam shifting cultivation is practised by only two hills district, Karbi-Anglong and Dima-Hasao.

Table 4.1 Area under shifting cultivation for 2015–16 in km^2

State	Area under current jhum (2000)	Shifting cultivation area (2010)	Area under current jhum (2015–16)	Change in 2010 compared to 2000	Change in 2015–16 compared to 2000	% of TCA
1	2	3	4	5	6	7
Arunachal Pradesh	3088.08	1531.46	509.11	– 1556.6	– 2579.0	16.91
Assam	8391.48	239.56	52.6	– 8151.9	– 8338.9	0.13
Manipur	12,014.06	852.20	499.98	– 11,161.9	– 11,514.1	11.44
Meghalaya	2086.77	448.99	237.87	– 1637.8	– 1848.9	7.85
Mizoram	3761.23	3761.23	691.55	0.0	– 3069.7	36.78
Nagaland	5224.65	5224.65	979.33	0.0	– 4245.3	19.43
Tripura	400.88	254.11	30.62	– 146.8	– 370.3	0.63

Source (1) Department of Land Resources, Wasteland Atlas of India 2019, (2) Indian Council of Forestry Research and Education and published under Statistical Year Book-2014 by MoSPI. TCA = Total Cropped Area

The primary issue related to shifting cultivation practices is the reduction in 'jhum cycle' in the process of cultivation. As a result of the reduction in the 'Jhum Cycle', it has adversely affected economic yield with a gradual decline in profit over some time (Ramakrishnan, 1996). The data in Table 4.2 compares the financial output/input analysis of the jhum under different cycles in the Northeastern region. The reduction in overall economic return is pronounced when the cost is calculated based on prevailing markets rates. The decrease in financial recovery under the short cycle of jhum is related to reduced soil fertility and increased weed potential of the site. Though the study belongs much early period by Ramakrishnan (1996), it gives a fair assessment of the trend of economic return as 'Jhum Cycle' becomes shorter.

Another most essential effect of the shifting cultivation is that it leads to environmental degradation and disturbs the fragile ecosystem in the region. This happens as a consequence of the degradation of the natural forest in the region. As a result of the environmental anomaly, monsoon rainfall has been distributed unevenly, causing drought and heavy rainfall, which has resulted in flooding in the area. In Meghalaya, unabated jhumming has turned the once thick evergreen forest belt of Cherrapunji, a place that used to record the highest rainfall in the world into a dry brown scar. The tradition of jhumming has resulted in significant environmental variations in the province, as well as the extinction of endangered plants and animals. The Cherrapunji ecosystem that now stands desertified due to deforestation inflicted sometime in the distant past now refused to recover its original state. Linked with this drastic loss in biological diversity, human suffering is now immense. Water is a scarce commodity during the dry months in this high rainfall spot of the world, a contender for becoming the driest sport near Mousenram in the Khasi Hills district of Meghalaya. The tribal who is traditionally bound to the land and forest have been forced to seek other avenues for survival (Ramakrishnan, 1996).

Climate change is emerging as a major threat to livelihood and food security of millions of people in many places of the world (IPCC, 2014). The World Meteorological Organisation and the United Nations Environment Programme together founded the IPCC in 1988. The IPCC's mission is to evaluate data from scientific and technical literature pertaining to all important aspects of the climate change problem. Climate change has directly affect the agriculture sector, especially in developing countries whose economy depend mainly on agriculture and it indirectly affect economic growth, income distribution, household welfare and agricultural demand

Table 4.2 Monetary input–output under different jhum cycles in the North Eastern region (Rs./hect./yrs.)

Jhum cycle			
	30 yrs.	10 yrs.	5 yrs.
Input	2616	1830	896
Output	5586	3354	1690
Net gain/loss	2970	1524	794
Output/input	2.13	1.83	1.88

Source Ramakrishnan (1996: 25)

(Schmidhuber & Tubiello, 2007). It is a significant factor in the current situation for planning and management of water resources. Global Circulation Models (GCMs) are tools that are used in the simulation of the present and future climate changes. These are numerical models that represent the various physical processes of the earth–atmosphere–ocean system (Prudhomme et al., 2003; Wilby & Wigley, 1997).

4.5 Soil Erosion and Land Degradation in the Northeastern Region

Soil fertility plays an important role in agriculture in that it determines the agricultural productivity. Soil fertility reflects the presence of nutrients in the soil systems and the ability of the soil to supply these nutrients to crops on a sustainable basis (Swift & Palm, 2000). While there is an emphasis on agricultural productivity, soil fertility is directly dependent on the land use mechanisms. Soil erosion is a diffuse procedure varying spatially over a typical landscape. Soil erosion is caused by detachment and removal of soil particles from land surface. It is a natural physical phenomenon which helped in shaping the present form of the earth's surface (Das, 2002). Direct estimation of soil erosion at many points across a region is impractical. Physically, it is difficult to measure decline and to analyse the variation of climate need at least ten years of data to obtain an accurate measure of average annual erosion. Various scientists and researcher used the erosion prediction models to analyse the impact of decline on agriculture, off-site sedimentation and many more. Proper management within an area needs to analyse the magnitude of soil erosion that strategies help to understand what is considered acceptable and the effects of different conservation. It was experienced that soil erosion is one of the major environmental hazards in today's life. As per the study Narayan and Babu (1983), had analysed that in India 5344 m-tonnes of soil is being detached over the years. Huge soil loss is not only for the reduced storage reservoir capacity but also a reduction in nutrients laden agricultural land.

The loss of forest and green cover in the hill region is a significant adverse effect of shifting agriculture and as a result it causes heavy soil erosion and that further led to heavy water flow causing floods in the plain area of the valley of Brahmaputra and Barak. In addition, the loss of top soil in the hill regions, which accounts for almost 22% of the total soil due to shifting cultivation, causes a significant drop in soil fertility and reduction in production and further adversely affects the economic condition of the jhumias. It is a concern on the sustainability of the changing cultivation.

Climate change is also associated with various influencing factors such as heavy rainfall and various extreme events. India will experience more seasonal variation in temperature with more warming in the winters than in the summers (Kumar & Gautam, 2014). The average temperature in India will also increase up to 2.33–4.78 °C with a doubling in CO_2 concentrations at the end of this century, and this will lead to higher variability in country's rainfall, ultimately affecting the agriculture sector (Nagargade et al., 2017). Crops are susceptible to the changes in temperature and precipitation, and to rising atmospheric CO_2 concentration (Zhao et al., 2017). Rainfall erosivity factor, hydrological and erosive processes are used to examine the cycles of severe and violet rainfall, which can be used to reflect soil and water management activities in agriculture's climate change adaptation strategy. With the help of soil erosion forecast models, rainfall erosivity explains the erosive mechanism and conservation steps. It is a fact that it is difficult to measure soil erosion at a large scale; models are needed to examine soil loss by water erosion at large scale and global scale. A significant research gap in assessing soil processes such as water erosion has been described as a result of recent advances in soil forecasting and climate change (Panagos et al., 2017).

In the present study, we have examined at detailed estimates of soil degradation and their consequences over time. For this analysis, Global erosivity dataset from long-term high-resolution data was used. The environmental variables required to construct a global rainfall erosivity map and the pattern of heavy rain events are included in this global rainfall erosivity station dataset. Further, analysing the Spatio-temporal variations of land use a land cover over the time period were analysed using MODIS (Moderate Resolution Imaging Spectroradiometer) data. Then Global-scale soil erosion datasets were used to analyse their impacts on land use land cover.

The present study represents the eight states of Northeast India, i.e., Arunachal Pradesh, Assam, Manipur, Meghalaya, Mizoram, Nagaland, Sikkim and Tripura, popularly known as seven sisters and a cousin brother. Geographically it lies between 200–300 N latitude and 900–980 E longitude. Northeast is almost covering 8% of the countries landmass. The total geographical area of the Northeast region is 260,464.96 km^2 whereas 82,023.38 km^2 is covered by Arunachal Pradesh state, 78,431.03 km^2 is covered by Assam state, 22,308.22 km^2 is covered Manipur state, 22,429.21 km^2 is covered by Meghalaya state, 21,120.55 km^2 is covered by Mizoram state, 16,596.05 km^2 is covered by Nagaland state, 7112.47 km^2 is covered Sikkim state, 10,444.03 km^2 is covered by Tripura state. NER contains more than one third of the country's total biodiversity and is considered one of the 18 hotspots in the world. The region is physiographically categorised into Eastern Himalaya, the Patkai and the Brahmaputra and the Barak valley plains. It is mainly covered by Brahmaputra-Barak river system and their tributaries. Kangchenjunga is the 3rd highest mountain peak in the world with an altitude of 8586 m that lies in between the state of Sikkim and adjacent country Nepal. In the Northeastern states, 85.78% of soil belong to the category of red soils, 8.74% alluvial soil, 1.98% hill soils and 2.83% belongs to others (for details refer to Table 4.3).

Table 4.3 State-wise area by major soil type, 2019 (in hectare)

States	Alluvial soils	Glaciers	Hill soils	Peat soils	Red soils	Rock outcrops	Terai soils	Others	Total
1	2	7	9	10					
Arunachal Pradesh	251.61	–	–	–	8004.46	–	–	118.23	8374.30
Assam	1204.68	–	–	–	6292.32	–	–	346.80	7843.80
Manipur	318.04	–	428.44	–	1432.40	3.90	–	49.92	2232.70
Meghalaya	167.62	–	–		2075.28	–	–		2242.90
Mizoram	107.30	–	–	–	1840.44	–	–	160.36	2108.10
Nagaland	–	–	–	–	1604.44	–	–	53.46	1657.90
Sikkim	–	101.40	–	–	523.16	–	78.34	6.70	709.60
Tripura	243.08	–	90.92	2.40	706.69	–	–	5.51	1048.60
NER total	2292.33 (8.74)	101.4 (0.39)	519.36 (1.98)	2.4 (0.01)	22,479.19 (85.74)	3.9 (0.01)	78.34 (0.30)	740.98 (2.83)	26,217.9 (100.0)

Source National Bureau of Soil Survey and Land Use Planning (ICAR) Nagpur, as compiled from EnviStats India 2020, Vol. I, Social Statistics Division, Ministry of Statistics and Programme Implementation, Govt. of India

4.5.1 Global Rainfall Erosivity Database

In general, collecting rainfall data, calculating the erosivity factor for each rainfall station, normalising the R-factor and regional interpolation of R-factor point values are all steps in the estimation of a global erosivity map. The Global Rainfall Erosivity Database (GloREDa) offers a European scale rainfall erosivity database with rainfall stations. Data with wide temporal resolution was collected from many countries to establish the dataset. GloREDa is the most robust global archive, containing data from as many stations as possible and wide temporal resolution.

The data is crucial for global and continental evaluations of water-induced soil erosion, risk of flooding, and natural disaster preparedness. Soil erosion by water is very uncertain that ranges from 20 to 200 Pg per year as per global estimated by the European Soil Data Centre (ESDAC). The ESDAC offers a global rainfall erosivity database on 3625 rainfall stations and about 60,000 years of rainfall data with high temporal resolution (1–60 min). The rainfall erosivity values for a single station were interpolated and an R-factor map was produced using the Gaussian process Regression model in this database. Thirty arc-seconds (~ 1 km) of the pixel size of the Global Erosivity rainfall data which measures in MJ mm ha^{-1} h^{-1} yr^{-1}. The data were plotted in Arc GIS environment (Table 4.4).

The Evaluation of Land Use Land Cover (LULC) monitoring from MODIS dataset using GIS Environment analysis was performed on the basis of analysing changes in Northeastern region of India. Hence spatial datasets have been used for LULC mapping. MODIS datasets were used to analyse the spatial changes from 2001 to

Table 4.4 Description of data used for global rainfall erosivity, 2012

Data	Year	Spatial resolution	Source
Global rainfall erosivity	2012	~ 1 km	https://esdac.jrc.ec.europa.eu/content/global-rainfall-erosivity

2019. The MODIS data are freely available and downloadable from the lpdaac.usgs.gov.

The Moderate-resolution Imaging Spectroradiometer (MODIS) was launched by the National Aeronautics and Space Administration (NASA) for a terra product in December 1999. For the classification LULC, MCD12Q1 data were used, which is available in 500 m spatial resolution. Hence to extract the information about LULC and its changes from 2001 to 2019, three consecutive techniques were applied in this study. Primarily the data were preprocessed and resampled using MRT Tool. The MODIS reprojection tool was developed to support MODIS data that are available in Hierarchical Data Format (HDF). MRT software facilitates the use of MODIS data by format conversion, projection, spectral and spatial Subsetting. Secondly, the data were subset for the study region on the basis of shapefile for the state, and the data were classified using the International Geosphere-Biosphere Programme (IGBP) legend. IGBP classification method was structured by University of Maryland classification scheme. Hence, on the basis of IGBP classification techniques, we analysed the LULC from 2001 to 2019. Furthermore, the final layout was prepared in a GIS environment, and results were analysed for 2001, 2007, 2012 and 2019 (refer Tables 4.5, 4.6 and 4.7).

To assess the human-induced soil erosion, data for the period 2000–2012 were used to analyse the change (refer Table 4.6). In the developing countries, soil erosion is mainly affected by LULC change. The Revised Universal Soil Loss (RULSE) model was used to establish global soil erosion results. RULSE is a kind of detachment restricted model in which soil erosion can be expressed. For erosion research, RULSE style models have been able to reduce a very complex method to a reasonably simple operation. These models are based on the concept of a dynamic system, which includes climate erosivity, soil erosivity, and other considerations. For the RULSE factors analysis, rainfall erosivity factor (R-factor), K, LS and P factors were used

Table 4.5 Description of data used for MCD12Q1, 2001–2019

Data	Year	Spatial resolution	Source
MCD12Q1	2001–2019	500 m	https://lpdaac.usgs.gov/products/mcd12q1v006/

Table 4.6 Description of data used for global soil erosion, 2001–12

Data	Year	Spatial resolution	Source
Global soil erosion	2001–2012	25 km	https://esdac.jrc.ec.europa.eu/content/global-soil-erosion

Table 4.7 Zone wise statistics of rainfall erosivity

State	Min	Max	Mean	Std
1	2	3	4	5
Arunachal Pradesh	279.15738	12,369.4	5400.701	4167.24
Assam	5489.7866	12,777.7	10,087.11	1575.08
Manipur	2522.8105	9872.7	5008.677	1739.51
Meghalaya	1847.9138	12,033.7	9141.175	2653.29
Mizoram	4433.436	9890.57	8280.477	1130.17
Nagaland	2443.9478	12,380.2	7089.485	2477.07
Sikkim	382.41925	10,224.3	3287.1	3128.49
Tripura	8068.1553	10,827.7	9162.968	390.451

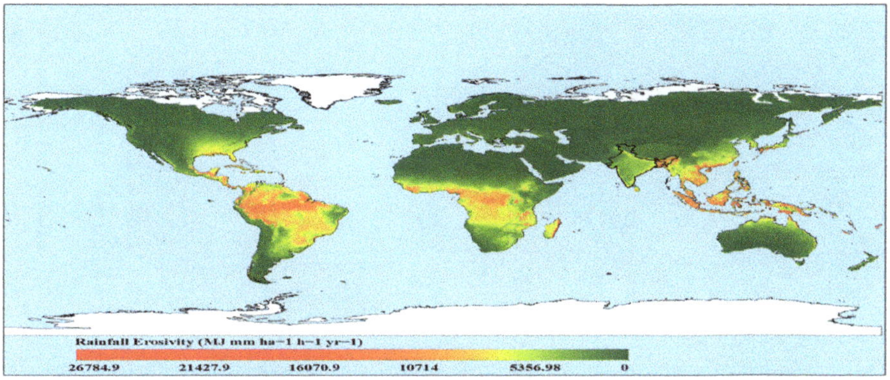

Chart 4.1 Global rainfall erosivity map (spatial resolution 30 arc-seconds)

to access the soil erodibility. Hence these data were further processed in Arc GIS environment.

GloREDa data offers erosivity values calculated using an R-factor from 3625 stations spread across 63 countries based on three components (a) Rainfall erosivity database at European Scale (REDES), (b) 1865 stations from 23 countries outside Europe and (c) 85 stations collected from the literature review. This information is derived from the compilation of high-resolution rainfall data from as many countries as possible. The R-factor values were interpolated using the Gaussian Process Regression (GPR) model to demonstrate the erosivity map. The global R-factor map's mean is 2190 MJ mm ha^{-1} h^{-1} yr^{-1}, with a standard deviation of 2974 MJ mm ha^{-1} h^{-1} yr^{-1}, indicating high heterogeneity. At the same time, the maximum of the global erosivity map is 26,784 MJ mm ha^{-1} h^{-1} yr^{-1}. For the North-eastern region of India, the erosivity map's mean is 7523 MJ mm ha^{-1} h^{-1} yr^{-1}, with a standard deviation of 3570 MJ mm ha^{-1} h^{-1} yr^{-1}, indicating high variability. At the same time, the maximum of the global erosivity map is 12,777 MJ mm ha^{-1} h^{-1} yr^{-1}. As per the figure of rainfall erosivity map, highest values are observed in the Assam, Meghalaya, Mizoram and Tripura whereas the lowest erosivity mainly located in Arunachal Pradesh, Nagaland, Manipur and Sikkim. In India the maximum of the erosivity has been mostly seen in the Northeastern region and the Eastern Ghats. In General, the High rainfall zone shows the high rainfall erosivity and vice-versa (refer Charts 4.1 and 4.2 and Table 4.7).

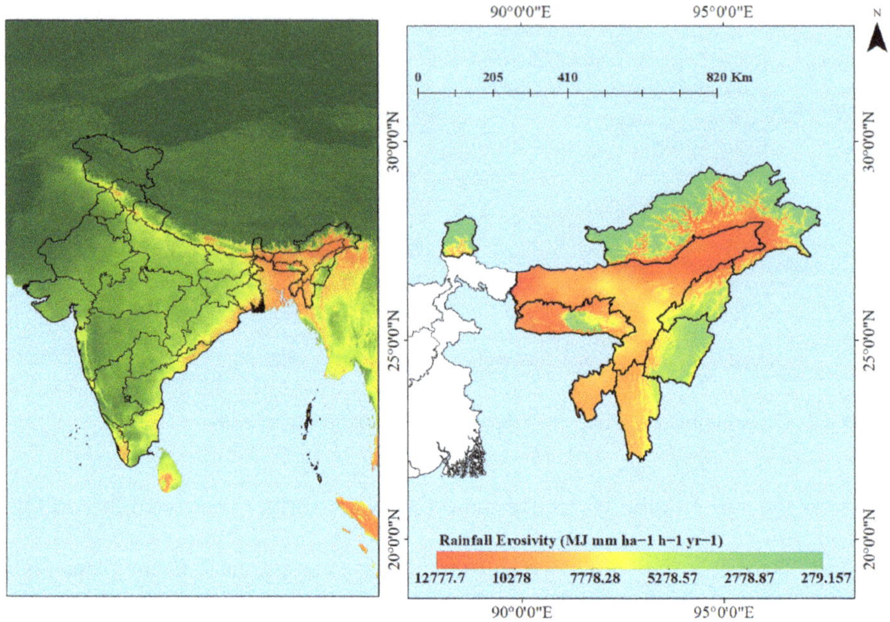

Chart 4.2 Rainfall erosivity map for the North Eastern region (spatial resolution 30 arc-seconds)

4.5.2 Land Use Land Cover (LULC) Change

The assessment of LULC change has been estimated in this study using the IGBP classification technique. The repetitive coverage of satellite data plays a significant role in this aspect by depicting the status of land use land cover. The LULC has been observed during the period of 2001–2019 for the entire Northeastern region of India. As per the statistical analysis, Forest cover areas, i.e. Evergreen Broadleaf forests are decreasing from 2001 to 2016; it increases from 2017 to 2019. In contrast, the Mixed forests and Evergreen Needleleaf forests have also increased over the time period. Woody Savannas and Closed shrublands have increased over the time period. Savannas are decreased from 2001 to 2019. Croplands have also increased during the period 2004–2013, afterwards it decreased. Urban and Built-up lands have also increased over the time period. Barren lands has also decreased over the time period (refer Charts 4.3, 4.4, 4.5, 4.6 and 4.7).

In this study, a detailed analysis has been carried out on the pattern of LULC change in different parts of the Northeastern region. The main features of LULC consisted of (i) Evergreen Needleleaf Forests, (ii) Evergreen Broadleaf Forests, (iii) Deciduous Needleleaf Forests, (iv) Deciduous Broadleaf Forests, (v) Mixed Forests, (vi) Closed Shrublands, (vii) Open Shrublands, (viii) Woody Savannas, (ix) Savannas, (x) Grasslands, (xi) Permanent Wetlands, (xii) Croplands, (xiii) Urban and Built-up Lands, (xiv) Cropland/Natural Vegetation, (xv) Permanent Snow and

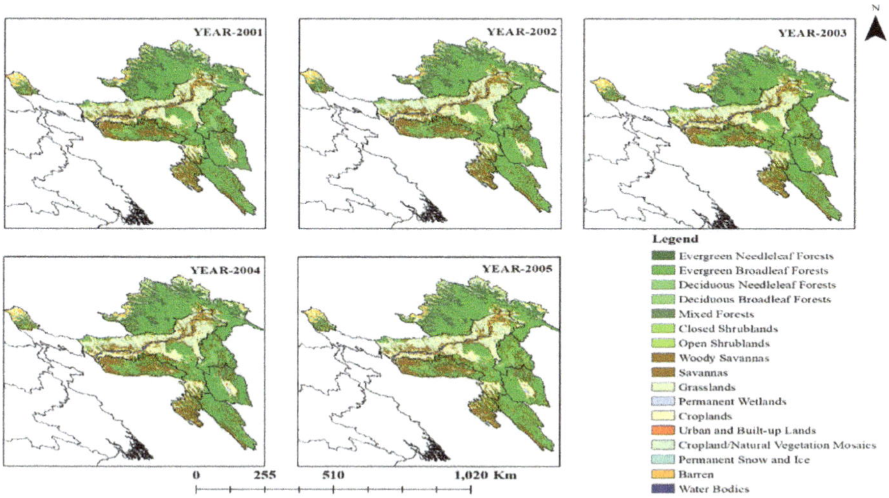

Chart 4.3 LULC change during 2001–2005 in the Northeastern region

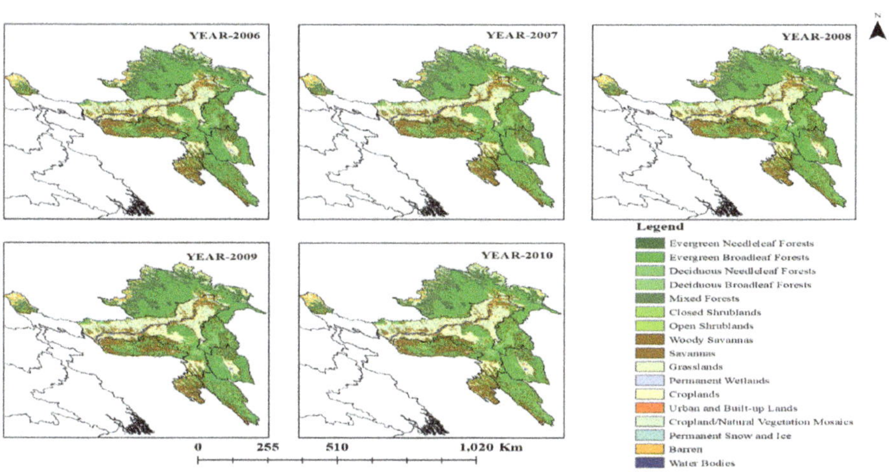

Chart 4.4 LULC change during 2006–2010 in the Northeastern region

Ice, (xvi) Barren and (xvii) Water Bodies. The analysis of Asset Account considering the component of shifting cultivation for the period 2011–12 as opening stock and 2015–16 as closing stock as published in EnviStat Vol. I, 2020 by Ministry of Statistics and Programme Implementation, Govt. of India have indicated a large percentage of shifting cultivation practices in all the states in the region. The state of Mizoram topped the list followed by Nagaland, Manipur, Arunachal Pradesh, Meghalaya and Tripura. The detailed analysis is reported in Table 4.8. Economic

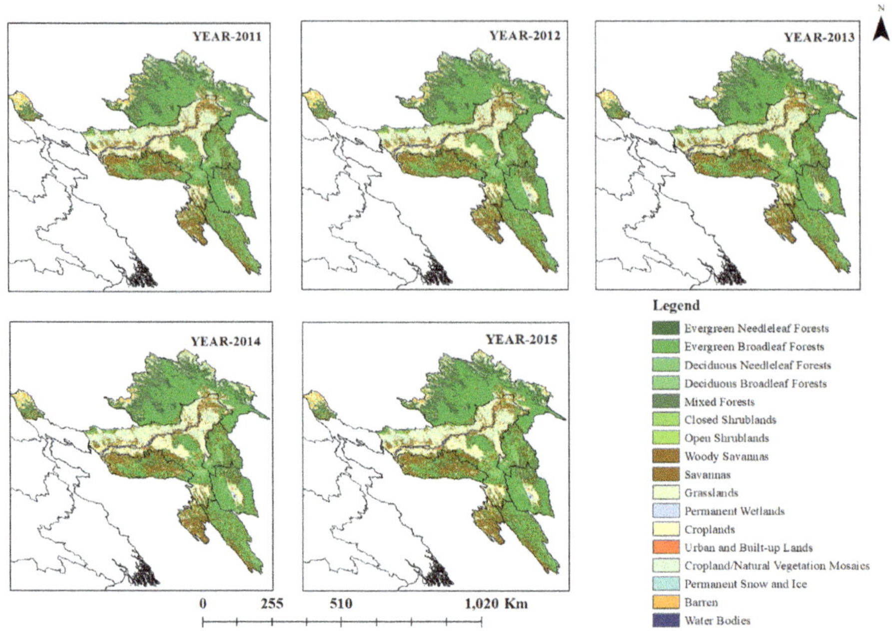

Chart 4.5 LULC change during 2011–2015 in the Northeastern region

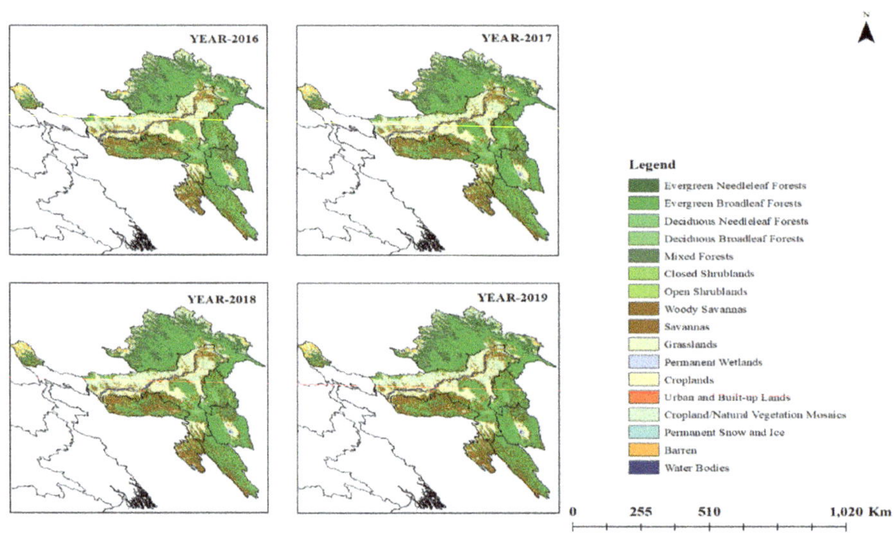

Chart 4.6 LULC change during 2016–2019 in the Northeastern region

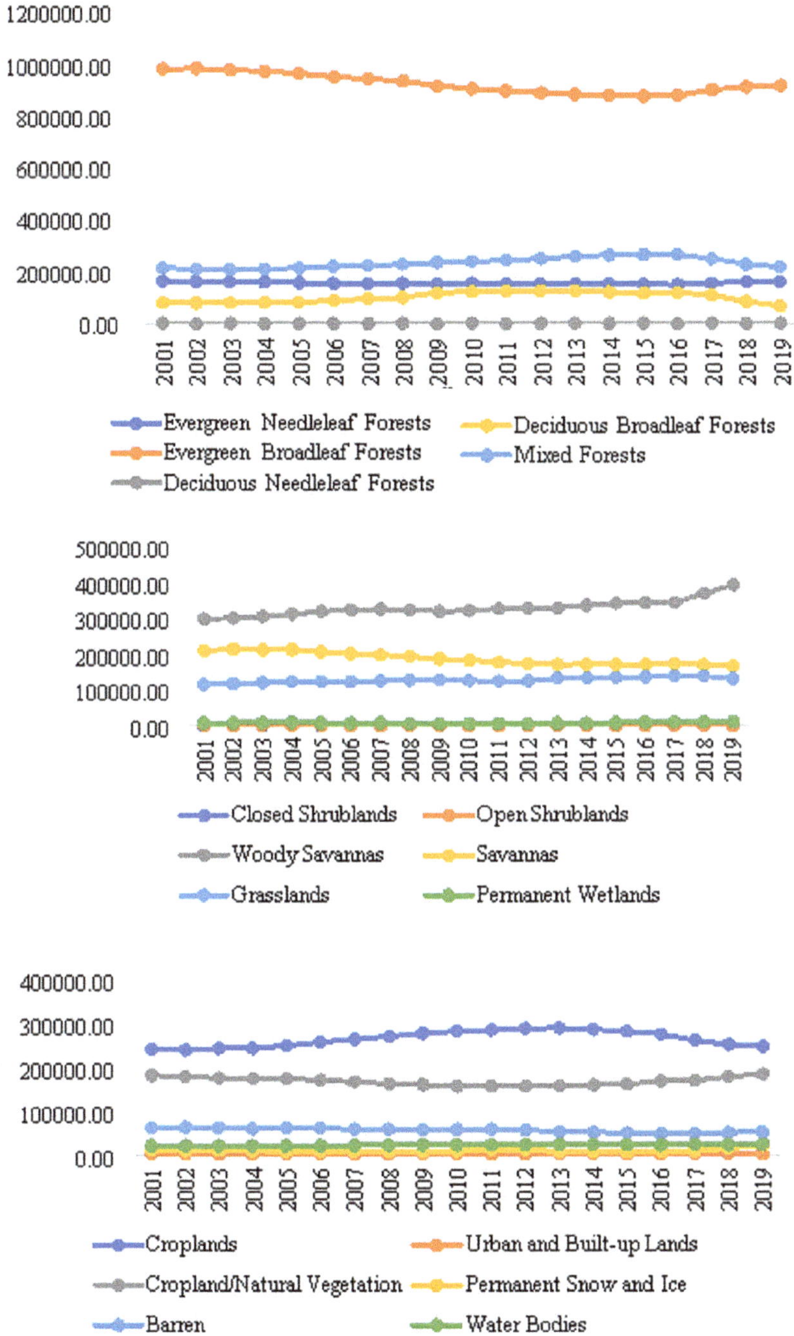

Chart 4.7 Graphical representation of LULC during 2001–2019

development, quality of life, natural resource conservation, and national food supply have also been affected by land use changes.

4.5.3 Soil Erosion

RULSE modelling based approach provides the potential rates of soil displacement by water erosion. The analysis was carried out for the period 2001 and 2012 for soil erosion. As per the analysis, the major soil displacement was seen in Assam in 2001, whereas the other areas were associated with less affected by soil erosion in 2001. Some moderate affect on soil erosion is seen in Arunachal Pradesh and Sikkim. In 2012 the displacement of soil erosion in Assam has increased and some of the other states were also affected by soil erosion. Maximum value of Arunachal Pradesh has increased i.e. 200 Mg ha^{-1} yr^{-1} in 2001 and 210 Mg ha^{-1} yr^{-1} in 2002 and also increased for Assam 47–48 Mg ha^{-1} yr^{-1} from 2001 to 2012 (refer Table 4.9 and Charts 4.8, 4.9).

The Global Rainfall Erosivity provides datasets at 30 arc-seconds (~ 1 km) spatial resolution. The Global Erosivity map is employed to measure soil erosion by water, flood risk, and natural disaster mitigation on an international and continental scale. The datasets are provided on the basis of RULSE model that provides a robust global rainfall erosivity factor to analyse or measure the erosive potential of rainfall. The results indicate the rainfall erosivity was very high in Assam, Meghalaya, Tripura and Mizoram whereas Arunachal Pradesh, Nagaland and Manipur have low and moderate rainfall erosivity. The analysis indicated that Assam has associated with high rainfall erosivity as well as increased soil erosion. Mizoram has also moderate to high soil erosion, whereas rainfall erosivity is also high.

The extent of soil erosion is specifically affected by the spatial trends of land use transition over time, which can impair land fertility and food security, ecological diversity, and carbon cycling. The Global soil erosion datasets were analysed at 25 km^2 spatial resolutions. The MODIS based MCD12Q1 datasets were used to analyse the LULC observation in Northeastern regions found high variation in Evergreen Broadleaf forests, savannas, cropland, urban Built-up lands, and water bodies. The analysis represents the decrease in croplands during the period 2001–2019. The results also represent that the croplands has increased during the period 2001–2010 after that it declined. Urbanisation and Built-up land were also increasing over the time period. A huge decrease in barren land was observed during the period 2001–2019 followed by tiny fluctuations in soil erosion during 2001–2012. The state of Assam has received more rainfall that rain feeds the Brahmaputra river. Assam experienced recurring flood and soil erosion every year which causes huge loss of life and property, agriculture land being destroyed and unbalances the socio-economic conditions of the people. The result clearly represents the high rainfall erosivity in Assam whereas soil erosion impact is also high.

Increased urbanisation, as well as changes in human related LULC, is major contributors to soil degradation, which has significant implications for nitrogen and

Table 4.8 State-wise asset account for shifting cultivation in land use land cover (area in km^2)

State	Opening stock (2011–12)	Addition to stock	Reduction in stock	Closing stock (2015–16)	Opening stock (% of respective state total)	Closing stock (% of respective state total)
1	2	3	4	5	6	7
Arunachal Pradesh	774.6	267.0	353.0	688.5	21.9	19.8
Assam	88.1	33.2	62.9	58.4	0.3	0.2
Manipur	461.8	488.3	311.3	638.8	22.0	24.6
Meghalaya	230.5	157.1	207.0	180.6	10.7	8.6
Mizoram	579.3	640.7	532.6	687.4	71.1	73.1
Nagaland	909.3	765.0	804.9	869.4	54.1	52.2
Sikkim	0.0	0.0	0.0	0.0	0.0	0.0
Tripura	16.4	29.0	14.8	30.7	0.7	1.4
NER total					7.3	7.4

Source Compiled from EnviStats India 2020, Vol. I, Social Statistics Division, Ministry of Statistics and Programme Implementation, Govt. of India

Table 4.9 Zone wise statistics of soil erosion during 2001 and 2012

	Min		Max		Mean		Std	
1	2	3	4	5	6	7	8	9
State name	2001	2012	2001	2012	2001	2012	2001	2012
Arunachal Pradesh	0.02	0.02	200.00	210.00	3.74	3.79	15.96	15.99
Assam	0.02	0.02	47.50	48.50	10.34	10.11	10.54	10.52
Manipur	0.02	0.02	130.54	130.54	5.90	5.90	22.11	22.11
Meghalaya	0.04	0.04	19.76	19.75	2.43	2.56	4.19	4.20
Mizoram	0.02	0.02	200.00	200.00	6.15	6.38	30.45	30.44
Nagaland	0.08	0.08	200.00	200.00	7.35	7.35	32.66	32.66
Sikkim	0.31	0.31	200.00	200.00	38.30	38.30	76.30	76.30
Tripura	0.01	0.01	18.33	18.33	2.94	3.12	4.81	4.78

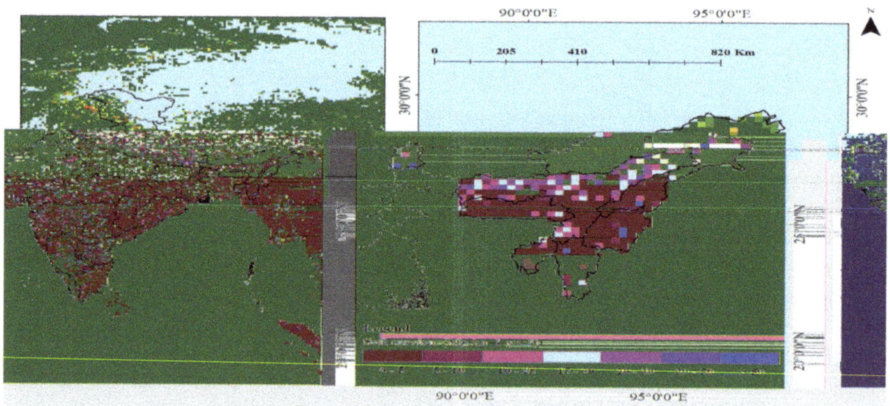

Chart 4.8 Soil erosion in Northeastern region of India, 2001

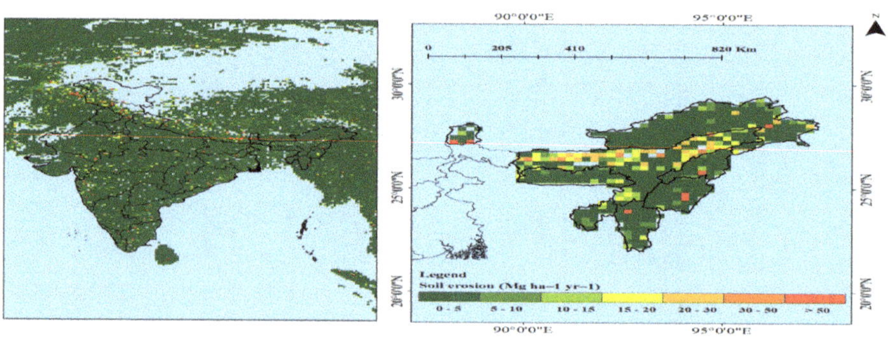

Chart 4.9 Soil erosion in Northeast region of India, 2012

carbon cycling, crop production, and other factors. In this study a combination of Remote Sensing and GIS has been used to analyse high resolution global potential soil erosion data and global rainfall erosivity data. The vulnerability of the Earth's surface to the energetic contribution of runoff is a key factor of water erosion management. In comparison, the consequences behind the water erosion mainly caused by soil degradation as indicated by the global pattern of rainfall erosivity. In contrast with these factors, LULC changes are analysed during 2001–2019. The results clearly shows that Assam has high soil erosion impacts, as well as the state also influence with rainfall erosivity. The land use change observes the big changes in Built-up land, barren land, cropland etc. The lowest change was observed in Arunachal Pradesh in all the factors.

Another very important aspect of environmental change is Land degradation, which is the loss of biodiversity and productivity that arises from the physical, chemical and biological degradation of the land. It affects the entire natural environment, resulting in losses of ecosystem services that have far-reaching effects on human welfare and the global economy. It also has a close connection with other major global issues, particularly climate change and biodiversity. Degraded land is a threat multiplier for communities, as it reduces people's ability to use their land and limits their access to resources (Borrelli et al., 2017). The main anthropogenic factors contributing to land degradation include deforestation and land clearing for economic use and to cope with increasing urbanisation. In some instances, the economic demand for agricultural land has led not only to land clearing but also over-cultivation, overgrazing, insufficient crop rotations and the overuse of agro-chemicals. These practices are focused on short-term production and profitability in order to meet the demand of growing populations. There is an urgent need to stop and reverse the process of land degradation (EnviStat, 2020). In the North-eastern Region, 16.34% of total land area is being estimated to be degraded land as compared to 2.69% for all India. The highest degraded land as a proportion of total area is in Mizoram (29.3%) followed by Nagaland (26.62%), Manipur (23.1%), Sikkim (17.4%) and Tripura (15.53%). No estimate for the state of Assam is available. The detailed account on state-wise degraded land is reported in Table 4.10. The EnviStat 2020 report, Government of India has also reported category-wise distribution of land degradation in different states of the Northeastern Region and also all India level data. It is revealed from the analysis of the data in Table 4.11 that in Northeastern region degraded land belong to the acidification was 59.9% and topped the list in 2005–06, which declined marginally to 57.1% in 2015–16. This was followed by water erosion, and it consisted of 26.5% of degraded land in 2005–06 and increased to 29% in 2015–16. The quantity of degraded land as a percentage of the total geographical area was 17.9% in 2005–06, which increased to 18.8% in 2015–16. For the country as a whole water erosion category of land degradation consisted of 55.9% in 2005–06 and remained stable at the same level in 2015–16. The second-highest category of degradation was wind erosion, and it consisted of 15.8% in 2005–06 and 15.7% in 2015–16. The degraded land for the country as a whole as proportion of geographical area was 17.9% in 2005–06 which increased to 18.8% in 2015–16.

Table 4.10 State wise district wise degraded land (area in hectare)

States	District	Total area	Total degraded land area	% degraded land area
1	2	3	4	5
Manipur	1. East Imphal	57,800	10,238	17.71
	2. West Imphal	51,900	15,098	29.09
	Total of affected districts	109,700	25,336	23.10
	State total	2,232,700	25,336	1.13
Meghalaya	1. East Garohills	260,300	34,201	13.14
	2. Jaintia Hills	381,900	178,666	46.78
	3. South Garohills	185,700	8003	4.31
	4. West Garohills	370,700	42,516	11.47
	Total of affected districts	1,198,600	263,386	21.97
	State total	2,242,900	263,386	11.74
Mizoram	1. Aizawl	357,631	109,184	30.53
	2. Champhai	318,583	184,795	58.01
	3. Kolasib	138,251	16,865	12.20
	4. Lawngtlai	199,119	95,965	48.19
	5. Lunglei	453,800	59,913	13.20
	6. Mamit	302,575	50,986	16.85
	7. Saiha	196,581	29,416	14.96
	8. Serchhip	142,160	70,702	49.73
	Total of affected districts	2,108,700	617,826	29.30
	State total	2,108,700	617,826	29.30
Nagaland	Kohima, Phek, Wokha, Zunheboto, Tuensang, Mokokchung, Mon	1,657,900	441,339	26.62
Sikkim	1. East	95,400	5922	6.21
	2. West	116,600	17,274	14.81
	3. North	422,600	94,963	22.47
	4. South	75,000	5323	7.10
	Total of affected districts	709,600	123,482	17.40
	State total	709,600	123,482	17.40
Tripura	1. West	303,300	21,385	7.05
	2. South	314,000	33,396	10.64
	3. North	210,070	60,732	28.91
	4. Dhalai	221,230	47,323	21.39
	Total of affected districts	1,048,600	162,836	15.53
	State total	1,048,600	162,836	15.53
NE total		10,000,400	1,634,205	16.34
All India		328,746,900	8,853,262	2.69

Source Soil and Land Use Survey of India, Ministry of Agriculture and Farmers' Welfare as compiled from in EnviStats India 2020, Vol. I, Social Statistics Division, Ministry of Statistics and Programme Implementation, Govt. of India

Table 4.11 Category-wise distribution of land degradation in different states of NER and all India level (in percentage)

State	Water erosion		Wind erosion		Water logging		Salinisation/ alkalisation		Acidification		Glacial		Anthropogenic		Others		Total	
	2005–06	2015–16	2005–06	2015–16	2005–06	2015–16	2005–06	2015–16	2005–06	2015–16	2005–06	2015–16	2005–06	2015–16	2005–06	2015–16	2005–06	2015–16
1	2	3	4	5	6	7	8	9	10	11	12	13	14	15	16	17	18	19
Arunachal Pradesh	42.2	45.1	0	0	3.7	3.6	0	0	36.1	34.3	0.1	0.1	0	0	17.6	17.0	100	100
Assam	50.6	48.4	0	0	34.9	36.2	0	0	10.6	10.0	0	0	0.9	1.6	3.7	3.8	100	100
Manipur	24.2	25.73	0	0	0.1	0.16	0	0	74.8	73.2	0	0	0	0	0.8	0.03	100	100
Meghalaya	11.9	13.8	0	0	0.05	0.05	0	0	73.6	72.9	0	0	1.1	1.2	13.4	13.1	100	100
Mizoram	27.8	36.2	0	0	0	0	0	0	72.2	63.8	0	0	0	0	0	0	100	100
Nagaland	3.0	4.6	0	0	0	0	0	0	97.1	95.3	0	0	0.1	0.1	0	0	100	100
Sikkim	7.0	7.0	0	0	0	0	0	0	0	0	14.8	14.8	0	0	78.2	78.2	100	100
Tripura	5.9	5.9	0	0	0.3	0.3	0	0	92.8	92.4	0	0	0.6	1.0	0.5	0.4	100	100
NER %LD	26.5	29.0	0	0	5.9	6.2	0	0	59.9	57.1	0.3	0.2	0.3	0.6	7.1	6.9	100	100
NER %TGA	4.7	5.4	0	0	1.1	1.2	0	0	10.7	10.7	0.05	0.05	0.06	0.1	1.28	1.3	17.9	18.8
All India %LD	55.9	55.9	15.8	15.7	2.0	2.0	7.1	7.1	3.3	3.3	3.2	3.2	0.5	0.7	12.2	12.1	100	100
All India %TGA	15.5	15.5	4.4	4.4	0.6	0.6	2.0	2.0	0.9	0.9	0.9	0.9	0.2	0.2	3.4	3.4	27.8	27.8

Source National Remote Sensing Centre, Government of India as compiled from EnviStats India 2020, Vol. I, Social Statistics Division, Ministry of Statistics and Programme Implementation, Govt. of India
TGA total geographical area; *LD* land degradation

References

Borrelli, P., Robinson, D. A., Fleischer, L. R., Lugato, E., Ballabio, C., Alewell, C., Meusburger, K., Modugno, S., Schutt, B., Ferro, V., Bagarello, V., Van Oost, K., Montanarella, L., & Panagos, P. (2017). An assessment of the global impact of 21st century land use change on soil erosion. *Nature Communications, 8*(1), Art. No. 2013.

Borthakur, D. N. (1992). *Agriculture of the North Eastern Region, Beece Prakashan, Guwahati* (pp. 137–151).

Bose, S. (Ed.). (1991). *Shifting cultivation in India.* Anthropological Survey of India.

Das, G. (2002). *Hydrology and soil conservation engineering practice.* Hall India.

Gadgil, M., & Guha, R. (1992). *The fissure land: An ecological history of India.* Oxford University Press.

IPCC. (2014). Summary for policymakers. Climate change 2014: Impacts, adaptation, and vulnerability. Part A: Global and sectoral aspects. In *Contribution of working group II to the fifth assessment report of the Intergovernmental Panel on Climate Change* (pp. 1–32). Cambridge University Press.

Kumar, R., & Gautam, H. R. (2014). Climate change and its impact on agricultural productivity in India. *Journal of Climatology & Weather Forecasting.*

Nagargade, M., Tyagi, V., & Kumar, M. (2017). Climate smart agriculture: An option for changing climatic situation. In S. Jurić (Ed.), *Plant engineering* (pp. 143–165). IntechOpen.

Narayan, D. V. V., & Babu, R. (1983). Estimation of soil erosion in India. *Journal of Irrigation and Drainage Engineering, 109*(4), 419–431.

Panagos, P., Borrelli, P., Meusburger, K., Yu, B., Klik, A., Lim, K. J., Yang, J. E., Ni, J., Miao, C., Chattopadhyay, N., Sadeghi, S. H., Hazbavi, Z., Zabihi, M., Larionov, G. A., Krasnov, S. F., Garobets, A., Levi, Y., Erpul, G., Birkel, C., ... Ballabio, C. (2017). Global rainfall erosivity assessment based on high-temporal resolution rainfall records. *Scientific Reports, 7*, 4175. https://doi.org/10.1038/s41598-017-04282-8

Prudhomme, C., Jakob, D., & Svensson, C. (2003). Uncertainty and climate change impact on the flood regime of small UK catchments. *Journal of Hydrology, 277*, 1–23.

Ramakrishnan, P. S. (1996). *Shifting agriculture and sustainable development.* UNESCO, The Parthenon Publishing Group.

Schmidhuber, J., & Tubiello, F. N. (2007). Global food security under climate change. *Proceedings of the National Academy of Sciences of the United States of America, 104*(50), 19703–19708. https://doi.org/10.1073/pnas.0701976104

Swift, M. J., & Palm, C. A. (2000). Soil fertility as an ecosystem concept: A paradigm lost or regained? In *Accomplishments and changing paradigm towards the 21st century.*

Wilby, R. L., & Wigley, T. M. L. (1997). Downscaling general circulation model output: A review of methods and limitations. *Progress in Physical Geography, 21*, 530–548.

Zhao, C., Liu, B., Piao, S., Wang, X., Lobell, D. B., Huang, Y., Huang, M., Yao, Y., Bassu, S., Ciais, P., & Durand, J. L. (2017). Temperature increase reduces global yields of major crops in four independent estimates. *Proceedings of the National Academy of Sciences, 114*(35), 9326–9331.

Chapter 5
Socio Economic Analysis of Sample Households in Northeast India

5.1 Statewise Location of Sample Households

The field studies, in connection with the present study, have been conducted as per the sample design discussed in Chap. 1. Seven states in the region, viz., Mizoram, Tripura, Manipur, Meghalaya, Assam, Nagaland and Arunachal Pradesh were selected in connection with the study for obtaining sample data. The names of selected districts, sub-division, villages, number of sample household and sample population are depicted in Table 5.1.

The states of Mizoram and Tripura have been considered for sample survey covering all the districts in these states for an exhaustive analysis. From other states, two villages each have been selected for conducting a sample survey. A total of 1469 households have been selected for conducting a sample survey covering a population of 7067. In view of the coverage of all the districts for the state of Mizoram and Tripura, the sample size becomes 642 and 321 in both the states respectively. The names of state-wise sample villages are mentioned in Table 5.1.

5.2 Occupational Distribution Patten

The analysis of the occupational distribution of sample data reveals that about 79% of households of the sample data are engaged in shifting cultivation in the NER. However, there are wide variations across different states in the NER in terms of dependency pattern on shifting cultivation. Interestingly in the state of Meghalaya, which is a hilly state with more than 95% inhabited by the indigenous population, only 47.4% sampled households are dependent on shifting cultivation. This fact indicates switching over of rural people in other sources of livelihood. The transformation in rural livelihood is an optimistic scenario for the state of Meghalaya. In this state, about 24% people in the sample households are dependent on self-employment in

N. Roy et al., *Climate Change, Shifting Cultivation and Livelihood Vulnerabilities in India*, https://doi.org/10.1007/978-3-031-54927-4_5

Table 5.1 Number of sample farm households and farm population in the selected villages in Northeastern region

District	Sub-division	Village	Sample household	Sample population
(1)	(2)	(3)	(5)	(6)
Mizoram				
Serchhip	East Lungdar (part)	N. Mualcheng	42	229
		Lungchhuan	40	218
Aizawl	Aibawk	Sialsuk	40	229
	Darlawn	Ratu	40	209
Mamit	Reiek	Reiek	40	196
	West Phaileng	West Phaileng	40	194
Lunglei	Lunglei	Ramlaitui	40	199
		Thuampui	40	214
Lawngtlai	Lawngtlai	Chawngte 'P'	40	184
	Chawngte	Damdep II	40	215
Champhai	Khawzawl	Dulte	40	196
		Chhawntui	40	212
Kolasib	Bilkhawthlir	Saihapui K	40	208
	Tlangnuam (part)	Serkhan	40	219
Saiha	Saiha	Lower Theiva	40	137
		Tuisih	40	127
State total			642	3186
Tripura				
Sepahijala	Bishalgarh	Badiyar Dighi	20	75
		Routh Khala	20	97
Khowai	Teliamura	Billaihampara	20	71
		Tirthamani	20	88
North Tripura	Kanchanpur	Gourchandra para	20	77
	Kanchanpur	Murari para	20	82
Unakuti	Kailashahar	Hiracherra	20	98
	Kailashahar	Unakoti	20	88
West Tripura	Jirania	Kengrai	20	96
	Jirania	Patnipara	20	89
Dhalai	Chawmanu	Makarchhera	20	99
	Chawmanu	Manikpur	20	113
Gomati	Udaipur	North Barmura	20	84
	Udaipur	South Barmura	20	91
South Tripura	Sabroom	Purba Sabroom	21	129

(continued)

Table 5.1 (continued)

District	Sub-division	Village	Sample household	Sample population
(1)	(2)	(3)	(5)	(6)
	Sabroom	Sada Shing para	20	82
State total			321	1459
Manipur				
Kangpokpi	Saitu Gamphazol	C. Munnom	20	140
	Saitu	Kholep	20	114
Chura Chand Pur	Saikot	Saidan	20	89
		Kangjam	30	80
State total			90	423
Meghalaya				
Ri Bhoi	Nongpoh	Purduwa	20	136
		Marmain	26	133
West Jaintia	Jowai	Iewthymmei	30	154
		Saphai	25	144
State total			101	567
Assam				
Dima Hasao	Haflong	Borolarpheng	30	162
		Puralangso	30	154
Karbi Anglong		Hojaipur	20	59
		Sipahi Basti	30	76
State total			100	451
Nagaland				
Kohima	Sechii-Zubza	Khonoma	60	121
	Jakhama	Phesama	30	152
Phek	Phek	Tuzatsu	25	118
		Kotisu	21	106
State total			135	497
Arunachal Pradesh				
Lower Dibang Valley	Anini	Amuli	20	26
		Alinye	20	198
Kurum Kumey	Pulin	Langdung	20	155
		Sanglum	20	105
State total			80	484

Source Field Survey, 2019–20

Table 5.2 Occupational distribution of sample farmers in Northeastern states (in percentage)

Village	Shifting cultivator	Wage earners	Salary earners	Self-employment in non-farm	Others	Population below BPL category
1	2	3	4	5	6	7
Mizoram	95.0	5.8	5.8	6.2	6.2	66.3
Tripura	76.3	15.8	3.1	7.0	9.0	68.9
Manipur	65.5	2.2	17.8	0.0	11.0	14.4
Meghalaya	47.4	2.0	10.0	24.0	7.0	22.6
Assam	91.67	5.00	3.33	5.5	2.6	53.3
Nagaland	77.4	18.0	3.0	15.0	2.0	63.6
Arunachal Pradesh	98.8	6.2	48.5	5.0	8.4	93.6
NER average	78.9	7.9	13.4	8.6	6.6	54.7

Source Field Survey, 2019–20

non-farm activities. It is also to be noted that 54.7% of sample households are living below the poverty line (refer Table 5.2) and have little opportunity to engage to elsewhere. This finding indicates an underdeveloped scenario in sampled villages on the respective states. In Arunachal Pradesh, about 99% of sample households are dependent on shifting cultivation with 93% people living below the poverty line.

The relation between shifting cultivators and the number of people living below the poverty line in different states of the region, as observed from the sample data is depicted in Chart 5.1. Except in the state of Manipur, in all other states, higher percentages of shifting cultivators contributed to the people living below the poverty line. This scenario indicates the underdeveloped character of shifting cultivation in these states.

5.3 Pattern of Income in the Households

Table 5.3 shows the distribution of income earned by sample households from different sources. It is observed from these distribution patterns that about 79% of sample households are engaged in agriculture (refer Table 5.2) earning about 47% of income (refer Table 5.3). The analysis also reveals that 24% of income is earned by only 8.6% self-employed in non-farm activities. The highest percentage of income, i.e. 76%, was made from agriculture in the state of Nagaland and the lowest in Arunachal Pradesh. This finding substantiated the higher proportion of people living below the poverty line in Arunachal Pradesh. The highest proportion, i.e. 37.5% of income earned from the farm and non-farm wages in the sample households of the state of Assam and that of lowest, i.e. 2% in the state of Mizoram. The sample households has earned about 32% of the income from the salary in Arunachal Pradesh and

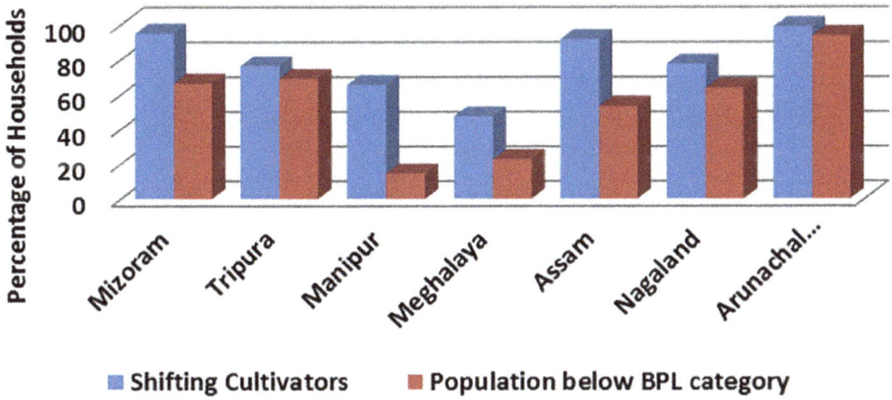

Chart 5.1 Shifting cultivators and population living below poverty line in sample household of Northeastern region (in percentage)

lowest, i.e. 3% in the state of Nagaland. In the case of self-employed in non-farm activities among the sample households, the state of Tripura was ahead of others. The families earned about 42% of income from self-employment activities in Tripura and the lowest figure, i.e. 11% in case of Assam. Thus, though agriculture dominates in terms of generating highest income but proportionately accommodates higher per cent of sample households for their livelihoods. It also indicates the underdeveloped nature of agriculture in all the states of the Northeastern region.

5.4 Cropping Pattern and Production in Sample Households

The data in Table 5.4 show the share of area under different crops production in all the sample states of the Northeastern region in the total cropped area of the respective state. It is observed that area under rice production has the highest share of the cropped area followed by maize, mustard and pulses for the region as a whole with wide variations across different states. For instance, for rice, about 89% area covered by the sample households in the state of Mizoram and lowest, i.e. 39.5% in the state of Meghalaya. The household level landholding pattern for rice cultivation shows 2.3 acres in the region as a whole. In comparison, the sample farmers in the state of Arunachal Pradesh allocate 3.0 acres for rice and that of lowest, i.e. 1.4 acres in Meghalaya. The cropping pattern in terms of area under major crops in depicted in Chart 5.2.

Table 5.3 Average yearly income from different sources (in rupees) in the sample households in Mizoram[a]

Village	Income source				
	Agriculture and allied activities	Farm and non-farm wages	Salary	Self-employed in non-farm activities	Total income
Mizoram	53,955 (52.0)	2194 (2.0)	13,047 (13.0)	33,620 (33.0)	100,878 (100.0)
Tripura	350,000 (45.2)	46,800 (6.0)	50,563 (6.5)	327,167 (42.2)	774,530 (100.0)
Manipur	63,486 (60.8)	5499 (5.3)	24,063 (23.0)	11,364 (10.9)	104,411 (100.0)
Meghalaya	169,228 (59.8)	14,528 (5.1)	26,417 (9.3)	72,586 (25.7)	282,759 (100.0)
Assam	202,345 (43.6)	174,000 (37.5)	36,000 (7.8)	52,000 (11.2)	464,345 (100.0)
Nagaland	287,531 (76.0)	8003 (3.0)	7238 (3.0)	45,325 (18.0)	348,097 (100.0)
Arunachal Pradesh	244,000 (28.1)	183,750 (21.2)	275,875 (31.8)	163,333 (18.8)	866,958 (100.0)
NER average	195,792.1 (46.6)	62,110.6 (14.8)	61,886.1 (14.7)	100,770.7 (24.0)	420,282.6 (100.0)

Source Field Survey, 2019–20

[a] Income in Rs and figures in the parentheses indicate respective percentages

Table 5.4 Area under different crops and their percentage share in cropped area in the sample households in Northeast India

Village	Crops							
	Rice		Maize		Pulses		Mustard	
	Area in acre	Percentage share	Area in acre	Percentage share	Area in acre	Percentage share	Area in acre	Percentage share
Mizoram	1.5	89.2	0.1	8.1	0.0	1.3	0.0	1.4
Tripura	1.9	49.6	0.7	16.6	0.6	14.5	0.8	19.4
Manipur	3.0	86.0	0.5	13.6	0.0	0.4	0.0	0.0
Meghalaya	1.4	39.5	0.2	5.4	0.0	0.0	1.9	55.1
Assam	1.6	67.4	0.7	28.1	0.04	1.8	0.07	2.8
Nagaland	2.3	65.7	0.5	14.9	0.2	7.0	0.4	12.3
Arunachal Pradesh	4.4	82.7	0.6	10.4	0.2	3.3	0.2	3.6
NER average	2.3	68.6	0.5	13.9	0.1	4.0	0.5	13.5

Source Field Survey, 2019–20

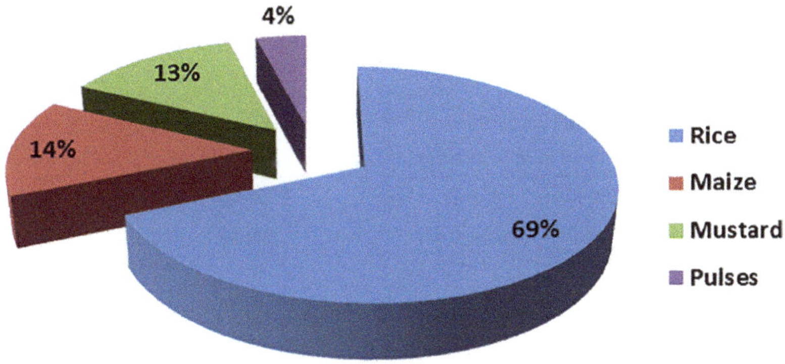

Chart 5.2 Cropping pattern in sample households in Northeastern region

The data in Table 5.5 revealed that 44.4% of the total cropped area is devoted to summer vegetables and that of 55.6% area in winter vegetables in the Northeastern region (also refer to Chart 5.3). The essential summer vegetables grown in the region are cucumbers, tomatoes, peppers, squash, beans, berries etc. mostly cultivated in rainfed conditions. The winter vegetables grown in the area consisted of cabbage, cauliflower, capsicum, spinach (palak), carrot, green peas etc. which require artificial watering to a great extent. The sample households cultivated the highest share of summer vegetables, i.e. 94.6% of cropped area in the state of Mizoram followed by Manipur (67%), Nagaland (56%), Assam (44%) and Tripura (37.3%). In thc states of Meghalaya and Arunachal Pradesh, the percentage share of summer vegetables was 3.2% and 9.0% respectively. There is no uniform pattern in the share of summer and winter vegetables across different states in the region. This finding indicates the suitability of a particular type of vegetables in different states due to physiographical and topographical conditions.

There is a wide variation in the operational holding size of sample households. The highest number of holding area lies between 2.5 and 5 acres, and it is also found that about 41% of the farm holding are within this size class for the Northeastern region as a whole. In the holding size class categories, 0.5–1.5 and 1.5–2.5 acres, about 38% of sample households existed across the states in the region (Table 5.6). Interestingly in the minimum size category of holding, i.e. 0.5–1.5 acres 301 farm households distributed across the region and this is the highest number followed by 299 farm households in 2.5–3.5 acres and 298 in 3.5–5.0 acres size of operational holdings. The distribution of farm households in different size classes is highly skewed across the region (refer to Chart 5.4).

Table 5.5 Area under seasonal vegetables and their percentage share in cropped area in the sample households in Northeastern region

Village	Crops			
	Summer vegetables		Winter vegetables	
	Area in acre	Percentage share	Area in acre	Percentage share
Mizoram	1085.0	94.6	62.0	5.4
Tripura	39.39	37.3	66.12	62.7
Manipur	89.0	67.0	46.0	33.0
Meghalaya	1.0	3.2	30.38	96.8
Assam	46.0	44.0	58.0	56.0
Nagaland	39.8	56.0	35.96	44.0
Arunachal Pradesh	3.0	9.0	29.0	91.0
Average NER	1303.2	44.4	327.5	55.6

Source Field Survey, 2019–20

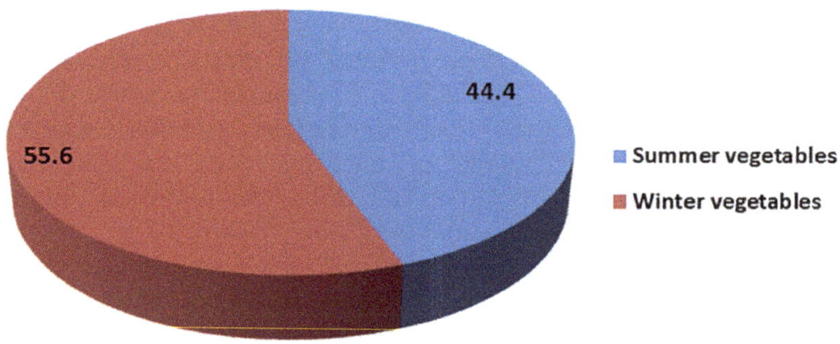

Chart 5.3 Percentage share in cropped area in the sample households in Northeastern region

Table 5.6 Distribution of sample farms according to size of operational holdings in the Northeastern region

State	Size class of operational holdings (in acre)						
	0.5–1.5	1.5–2.5	2.5–3.5	3.5–5.0	5.0–8.0	8.0–20.0	Total
Mizoram	108	116	127	116	102	73	642
Tripura	123	51	75	24	32	11	321
Manipur	14	1	25	45	1	4	90
Meghalaya	12	35	11	39	4	0	101
Assam	9	22	29	34	4	2	100
Nagaland	31	23	17	11	12	11	105
Arunachal Pradesh	4	5	12	27	21	11	80
Average NER	301 (20.9%)	253 (17.6%)	299 (20.8%)	298 (20.7%)	176 (12.2%)	112 (7.8%)	1439 (100%)

Source Field Survey, 2019–20

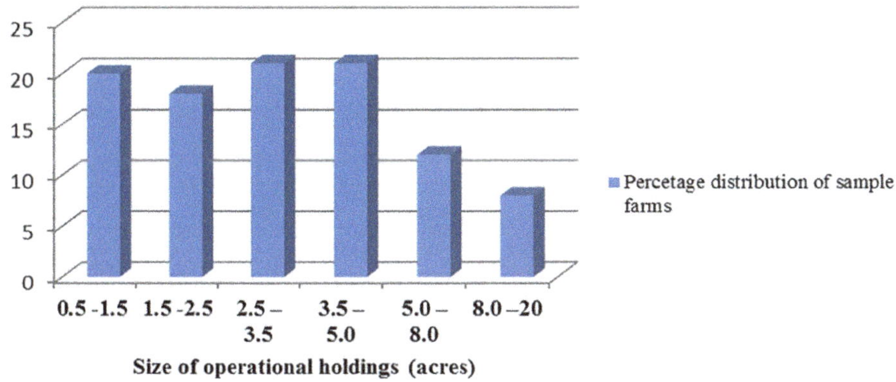

Chart 5.4 Distribution of sample farms according to size of operational holdings (in acres)

In Table 5.7, it is observed that in the Northeastern region as a whole, rice is the most produced crop. Rice is the staple food for the population across different states in this region. The yield of rice in the region estimated as 1199 kg per acre, which is much less than the country as a whole. It is to be noted that in shifting cultivation practice rice is generally grown without using any modern technology and nutrients. The highest yield of rice, i.e. 1560 kg per acre was obtained in the state of Assam followed by Mizoram (1400 kg), Meghalaya (1200 kg), Arunachal Pradesh (1171 kg), Nagaland (1127 kg), Manipur (1121 kg) and lowest in Tripura (919 kg). Apart from rice other principal crops as evident from the sample data are maize, pulses and mustard. The sample households cultivated other crops like Maize, Pulses and Mustard in a limited area among total cropped area in different states in the region. The production and yield of these crops were not encouraging.

In the survey, respondents were asked to share their opinion about the changes in cropping pattern in their own farm and also in the neighborhoods. Perception of farmers plays a very important role in taking decisions for crop planning and use of their farmlands for different cultivations across different seasons. Table 5.8 records the perception of villagers regarding variations in the pattern of cropping during the last ten years. Almost 62% of the sample farmers for the Northeastern region as a whole said that they had perceived changes in cropping pattern as a reaction to changes in climatic conditions. In this respect all respondents from Meghalaya followed by 94% in Mizoram, 85% in Manipur, 74% in Nagaland and 69% in Tripura believe that they perceive changes in the cropping pattern. However, 98% of farmers in Assam and 86% in Arunachal Pradesh did not perceive any change in the cropping pattern.

Table 5.7 Production of different crops and their percentage share in cropped area in the sample households in the Northeastern region

Village	Crops							
	Rice		Maize		Pulses		Mustard	
	Production in kg	Yield per acre	Production in kg	Yield per acre	Production in kg	Yield per acre	Production in kg	Yield per acre
Mizoram	1318.4	1400	56.6	0.4	3.0	53.0	5.2	48.9
Tripura	1462.0	919.1	182.9	322.3	190.3	223.1	80.9	84.3
Manipur	3608.5	1121	167.0	2.6	37.0	123.0	52.6	45.6
Meghalaya	4352.7	1200	756.2	34.9	118.6	98.0	345.0	70.2
Assam	2192.0	1560	342.6	231.0	134.0	89.1	178.0	84.3
Nagaland	2310.2	1127.4	366.3	550.4	104.1	241.8	248.1	428.3
Arunachal Pradesh	462.7	1071.8	9.4	189.2	22.2	330.0	462.7	1071.8
Average NER	2243.8	1199.9	268.7	190.1	87.0	165.4	196.1	261.9

Source Field Survey, 2019–20

Table 5.8 Perceived change in cropping pattern over last 10 years (in percentage)

Village	Yes	No	Total
Mizoram	93.9	6.1	100.0
Tripura	69.4	30.6	100.0
Manipur	85.5	14.5	100.0
Meghalaya	100.0	0.0	100.0
Assam	1.7	98.3	100.0
Nagaland	74.2	25.8	100.0
Arunachal Pradesh	14.0	86.0	100.0
Average NER	62.7	37.3	100.0

Source Field Survey, 2019–20

5.5 Awareness Level on Climate Change at the Household Level

The figures in Table 5.9 analysed the awareness of the households about climate change, asking a simple question that if they have heard or know what changes in climate mean. The respondents revealed that they have heard about climate changes from different sources. It is to be noted that more than 61% of the sample households in the region have started to come to know about climate change from their neighbours followed by 59.6% from television, 47.6% from newspapers, 38.3% from health workers and 32.2% from the weekly magazine (refer Chart 5.5). However, there are wide variations in respect of responses from the households in different states in the region. For instance, in the state of Assam, no sample household received information about climate change from the newspaper, weekly magazines and health workers. In contrast, all the respondents in the state of Meghalaya received information from the sources of radio, television, neighbour's and health workers.

Table 5.9 Awareness about climate change from different sources (in percentage) in the sample households in the Northeastern region

Village	Newspaper	Weekly magazine	Radio	Television	Neighbours	Health workers
Mizoram	76.0	23.0	12.0	44.0	66.0	7.0
Tripura	15.9	0.0	11.1	51.4	34.3	8.1
Manipur	33.8	29.2	53.8	42.5	76.3	29.2
Meghalaya	39.3	11.0	100.0	100.0	100.0	100.0
Assam	0.0	0.0	18.3	96.7	1.7	0.0
Nagaland	76.5	73.0	56.9	16.2	93.5	56.9
Arunachal Pradesh	91.7	88.9	66.7	66.7	57.2	66.7
Average NER	47.6	32.2	45.5	59.6	61.3	38.3

Source Field Survey, 2019–20

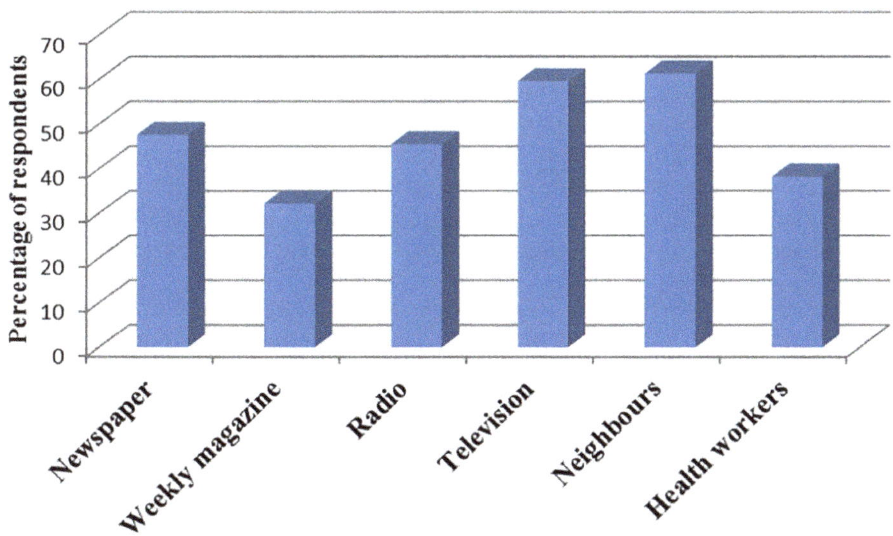

Chart 5.5 Awareness about climate change from different sources in farm households in Northeastern region

Interestingly the extension services by health workers for an awareness campaign about climate changes are widespread in the state of Meghalaya, Nagaland and Arunachal Pradesh. In the state of Manipur, 29% of respondents received information from health workers followed by 8% in Tripura, 7% in Mizoram and no respondents in Assam. Extension agencies form a very important part in an awareness campaign about climate change its consequences and remedial measures to be adopted by the shifting cultivators.

The data in Table 5.10 and Chart 5.6 reveals that about 67% of sample respondents indicated that they think the climate is gradually changing, while 9% did not answer for any specific opinion. Then we asked them about their opinion on specific forms on climate change like change in temperature, rainfall, wind, drought and humidity pattern. About 81% of the respondents in the region are of the opinion that there is an increase in temperature followed by 73% observed changes in rainfall, 51% answered changes in wind followed by 14% for drought occurred, and 38% commented for changes in humidity pattern. These observations on climate change are very pertinent in the context of their decision making for cropping pattern and adopting remedial measures.

To understand the causes of climate change as perceived by the respondents, a set of options were given to them to indicate their perceptions and understanding of climate change. Around 53% of the households identified deforestation as the main cause of climate change, while 23% perceived shifting cultivation as a reason behind climate change. About 46% perceived industrialisation/urbanisation as reasons for climate change, while 47% indicated automobile emission as the cause. Overpopulation as the reasons for climate change was indicated by 42% of sample respondents. However, there are wide variations in respect of perceived reasons, as indicated by the respondents across different states in the Northeastern region. For instance, in the state of Meghalaya, Nagaland and Arunachal Pradesh, only 5%, 8% and 8% of respondents indicated deforestation as the perceived cause of climate change. Similarly, in the state of Tripura, about 70% of respondents identified deforestation and overpopulation as the main reasons for climate change. From the analysis of the data, it is revealed that shifting cultivation was not the perceived reason for the respondents as the major reason for climate change (refers to Table 5.11 and Chart 5.7).

Table 5.10 Opinion about climate change in the sample households in the Northeastern states (in percentage)

	Changing	Not changing	Don't know	Total
Opinion about climate change	67.4	20.9	9.5	100.0
Change in temperature	81.5	7.1	11.4	100.0
Change in rainfall	73.7	23.5	3.3	100.0
Changes in wind	51.5	17.0	31.5	100.0
Drought occurred	14.0	23.7	68.0	100.0
Humidity pattern	38.1	12.9	49.0	100.0

Source Field Survey, 2019–20

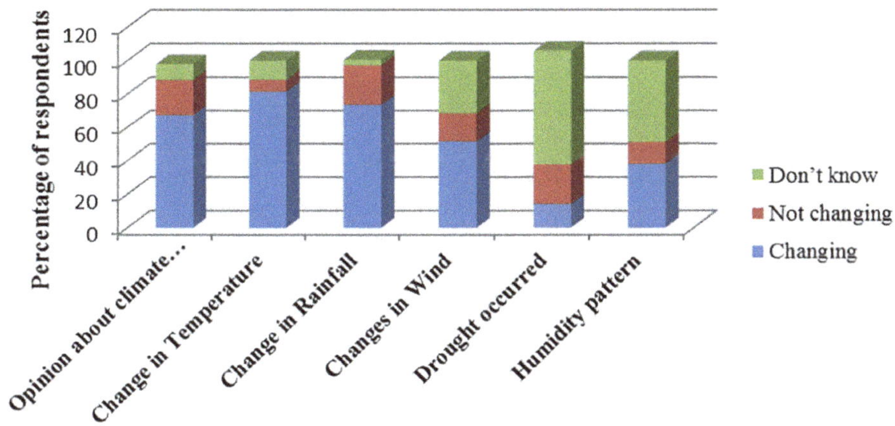

Chart 5.6 Opinion about climate change in the sample households

Table 5.11 Perceived reasons for climate change in the sample households in Mizoram (in percentage)

Village	Deforestation	Industrialization/ urbanisation	Automobile exhaust	Over population	Unscientific agricultural methods	Shifting cultivation
Mizoram	91.7	6.9	11.2	0.3	0.3	31.3
Tripura	69.2	1.3	2.2	71.7	9.7	10.9
Manipur	98.5	11.2	10.1	26.5	30.2	20.2
Meghalaya	5.7	76.8	84.0	89.3	100.0	15.6
Assam	95.0	90.0	71.7	13.3	10.0	31.0
Nagaland	8.3	76.5	86.2	64.3	75.4	33.3
Arunachal Pradesh	8.3	60.0	66.7	33.4	50.0	23.0
Average NER	53.8	46.1	47.4	42.7	39.4	23.6

Source Field Survey, 2019–20

5.6 Perception on Climate Change Mitigation

For any climate change study, one of the important aspects is the identification of the probable impact of climate change. In this respect perception of people plays a very important role in adaptation and mitigation strategies to be undertaken by them, and these contribute to society as a whole. Altogether ten impacts have been identified for inclusion in the questionnaire, and respondents were asked to indicate the impacts based on their perception. The analysis has been reported in Table 5.12 and Chart 5.8. Among the ten impacts, about 74% respondents indicated reduced vegetation as the important impact of climate change followed by soil erosion and

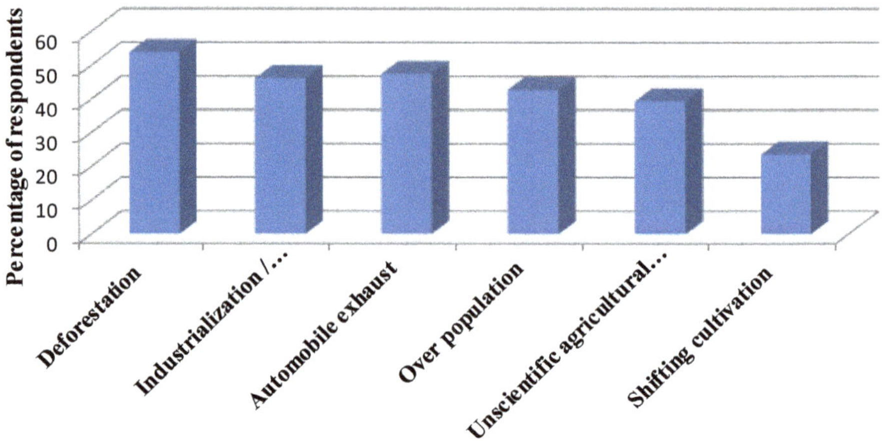

Chart 5.7 Perceived reasons for climate change in the sample households in Northeastern region

quality loss by 70% respondents, species loss by 66%, loss of soil organic content by 59% and the harsh environment by 52% respondents. Expansion of irrigated land and changes in vegetation type is considered as a consequence of climate change by 36% and 44% respectively. Forty one per cent sample households has considered climate change responsible in increased in vector-borne diseases, and 31% thinks it is the cause for reduced biological productivity. Increase in the forest fires has been perceived to be an outcome of change in climate by 40% sample households. This analysis thus identified several probable impacts of climate change perceived by the respondents across different states in the Northeastern region.

The impact of climate change led to livelihood vulnerabilities among the people in the society. Regarding vulnerability as perceived by the sample respondents, 3.5% referred it as highly vulnerable followed by 29.6% as vulnerable and 34.4% as less vulnerable. However, one third of the sample respondent's i.e. about 32.5% preferred not to answer this question (refer Table 5.13 and Chart 5.9). However, there are wide variations in respect of opinion expressed by different sample respondents in the region. This perception of sample households of the Northeastern region has important implications for their preparedness to face the challenges of the impact of climate change.

Although the sample respondents in the region are found to discuss climate conditions but only 15% of them have discussed the various adaptation measures that could be adopted as a safeguard against climate change, and 20.3% reported they did not discuss it ever. While 29.4% respondents indicated that they discuss adaptation strategies sometimes, a major percentage of them, i.e. 35.3% preferred not to answer this question. This analysis is reported in Table 5.14. Thus overall, there is very little discussion about adaptation measures to climate change across different states in the Northeastern region.

Table 5.12 Perception on impact of climate change in the sample households in the Northeastern region (in percentage)

Villages	Perceived reasons observed (%)									
	Reduced vegetation cover	Species loss	Changes in vegetation type	Reduced biological productivity	Expansion of irrigated land	Soil erosion and quality loss	Loss of soil organic content	Increased forest fires	Increase in vector borne diseases	Harsh environment
1	2	3	4	5	6	7	8	9	10	11
Mizoram	87.0	94.0	82.0	93.0	86.0	93.0	92.0	74.0	38.0	95.0
Tripura	81.6	79.6	63.3	8.9	12.2	35.4	29.2	2.5	0.0	33.3
Manipur	98.8	87.9	75.7	90.2	75.2	88.8	79.2	91.7	87.9	92.7
Meghalaya	6.6	85.8	75.9	27.4	43.1	98.9	100.0	100.0	100.0	50.0
Assam	98.3	0.0	0.0	0.0	0.0	25.0	0.0	0.0	0.0	0.0
Nagaland	84.9	82.4	12.4	0.0	42.1	84.9	82.4	12.4	0.0	42.1
Arunachal Pradesh	66.7	33.3	0.0	0.0	0.0	66.7	33.3	0.0	66.7	57.2
Average NER	74.8	66.1	44.2	31.4	36.9	70.4	59.4	40.1	41.8	52.9

Source Field Survey, 2019–20

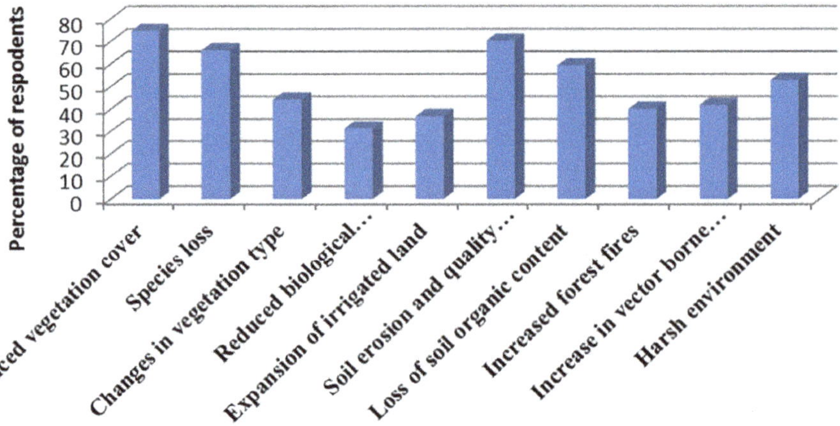

Chart 5.8 Perception on impact of climate change in the sample households in the Northeastern region

Table 5.13 Perception about vulnerability to climate change in the sample households in Northeastern region (in percentage)

States	Highly vulnerable	Vulnerable	Less vulnerable	No answer	Total
1	2	3	4	5	6
Mizoram	0.0	6.0	44.0	50.0	100.0
Tripura	0.6	43.1	33.4	22.8	100.0
Manipur	0.2	28.0	40.8	31.1	100.0
Meghalaya	0.2	35.2	16.9	47.7	100.0
Assam	1.0	23.0	37.0	39.0	100.0
Nagaland	15.9	42.8	29.2	12.1	100.0
Arunachal Pradesh	6.3	29.3	39.5	25.0	100.0
Average NER	3.5	29.6	34.4	32.5	100.0

Source Field Survey, 2019–20

An attempt has also been made in the present study to explore the proportion of respondents who have actually taken any measure to adapt the climate change mitigation (refer to Table 5.15). About 71% of the sample respondents have denied about taking any measures of adaptation to climate change. However, 28% of the respondents who acted passionately by adopting some mitigation measures is a positive reflection considering the remoteness of the sample respondents from different indigenous societies in the Northeastern region. Moreover, among the negative respondents, there is a chance that they did not understand what actually is meant by the term adaptation measure since most of the sample households are not educated.

For finding the involvement in using either crop or water or land management practices, questions were asked about the type of practices the sample respondents adopt

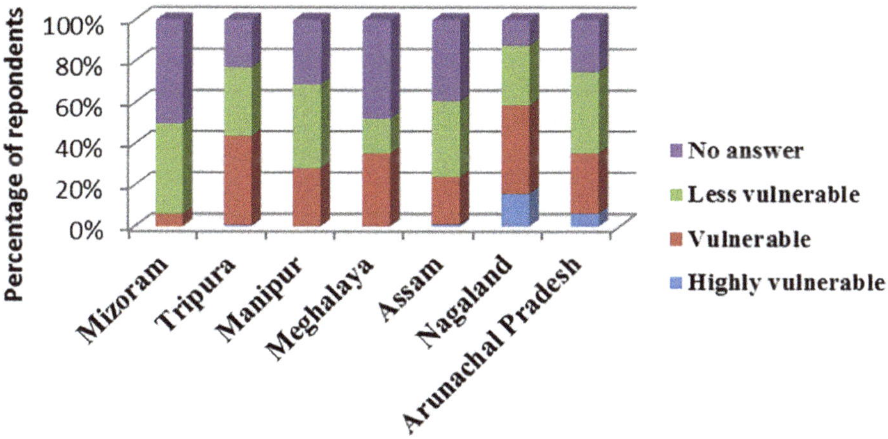

Chart 5.9 Perception about vulnerability to climate change in the Northeastern region

Table 5.14 Status of any discussion took place about climate change and adaptation measures in the sample households in the Northeastern region (in percentages)

States	Often	Sometimes	Not at all	No answer	Total
1	2	3	4	5	6
Mizoram	0	5	45	50	100.0
Tripura	0	3.5	59	37.5	100.0
Manipur	45	5	2	48	100.0
Meghalaya	31.0	23.0	8.0	38.0	10.0
Assam	2.0	56.0	2.0	40.0	100.0
Nagaland	20.9	54.5	19.4	5.2	100.0
Arunachal Pradesh	6.3	58.5	6.6	28.6	100.0
Average NER	15.0	29.4	20.3	35.3	100.0

Source Field Survey, 2019–20

Table 5.15 Measures adapted to climate change by the sample households in the Northeastern region (in percentage)

States	Yes	No	Total
1	2	3	4
Mizoram	5.0	95.0	100.0
Tripura	27.0	73.0	100.0
Manipur	48.7	51.3	100.0
Meghalaya	22.6	77.4	100.0
Assam	0.0	100.0	100.0
Nagaland	44.0	56.0	100.0
Arunachal Pradesh	54.0	46.0	100.0
Average NER	28.8	71.2	100.0

Source Field Survey, 2019–20

for specific measures. The responses received from them are depicted in Table 5.16. It is found that overall in the Northeastern region, 36.4% households have adopted any crop management measure, 33.5% taken measures of water management and 33.2% of them have taken land management measures. The question was also asked about diversification of income, and only 40.1% sample respondent has adopted measures related to income diversification whereas established rituals to protect the environment has been celebrated by 20.7%. However, there are wide variations across different states in the region in respect of the different practices adopted by them (refer Chart 5.10). For instance, a large majority of sample respondents practices all the listed measures of adaptation, including rituals. Interestingly 82% sample respondents from the state of Mizoram adopted rituals as one of the specific measures for adaptation to climate change. In case of the state of Tripura, a small percentage of respondents adopted all the listed adaptation measures. Only 11.9% adopted crop management followed by 7.8 of water management, 12.2% adopted land management, 24.8% using income diversification, and no respondents adopted rituals as mitigation practice for climate change. For the state of Assam, only 1% of respondents adopted crop management rituals for climate change adaptation measures. However, a large majority, i.e. 78% respondent's income divarication as an alternative to climate change adaptation measures.

In all other states like Manipur, Meghalaya, Nagaland and Arunachal Pradesh, all the practices adopted ranges between 10% and 50% by the respondents. It is to be noted here that large sample sizes collected from the states of Mizoram and Tripura compared to other states. Thus, shifting cultivators despite their awareness about climate change and its consequences are not adopting specific mitigation measures to a large extent. For understanding this paradoxical situation, it is necessary to identify the barriers for the adoption of specific standards to climate change by the

Table 5.16 Specific measures to adaptation taken by the sample households in the Northeastern region (in percentage)

Village	Specific adaptation measures				
	Crop management	Water management	Land management	Income diversification	Rituals
1	2	3	4	5	6
Mizoram	94.0	70.0	97.0	95.0	82.0
Tripura	11.9	7.8	12.2	24.8	0.0
Manipur	20.0	29.0	10.0	17.0	24.0
Meghalaya	49.0	45.0	58.0	47.0	12.0
Assam	1.0	45.0	12.0	78.0	1.0
Nagaland	41.6	29.5	30.9	14.9	17.8
Arunachal Pradesh	37.5	8.3	12.5	4.2	8.3
Average NER	36.4	33.5	33.2	40.1	20.7

Source Field Survey, 2019–20

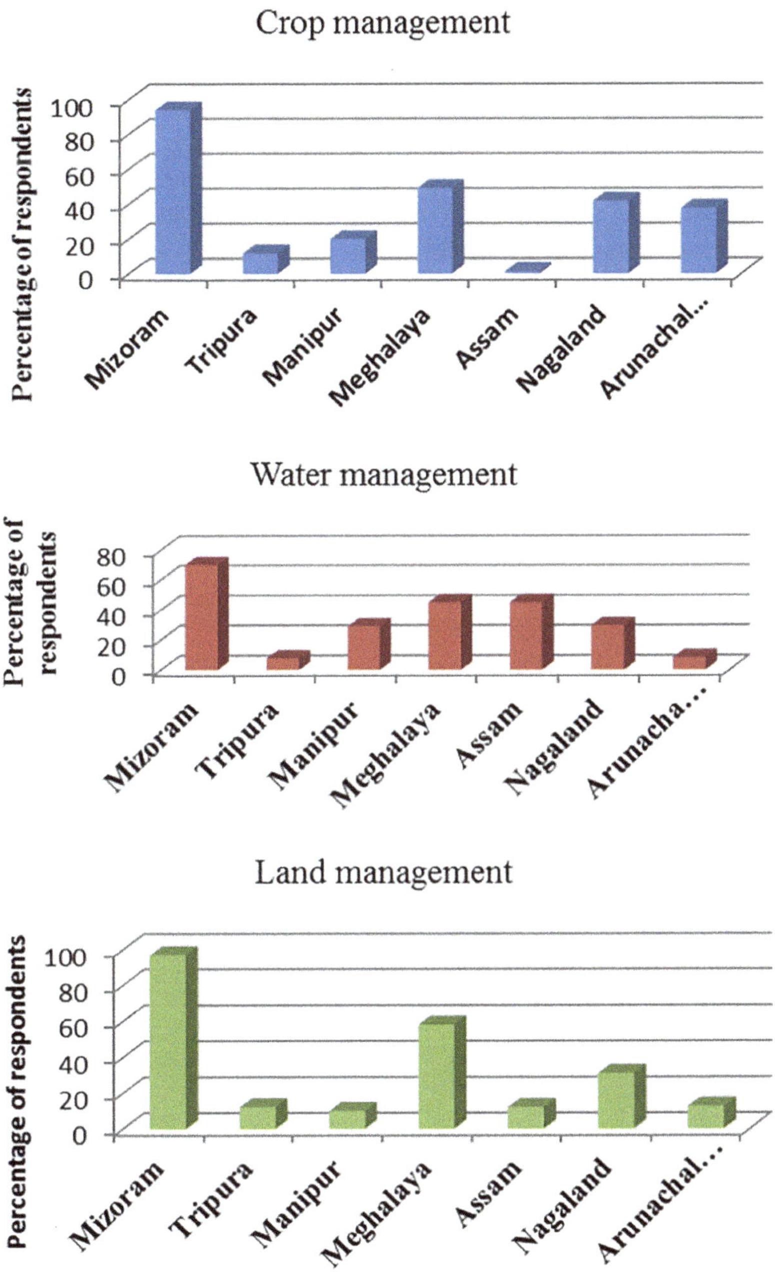

Chart 5.10 Specific measures to adaptation taken by the sample households in the Northeastern region

respondents. Mostly obstacles arise both from the absence or inefficiency of institutional and non-institutional factors. The identification of these barriers is crucial in the context of specific measures to be adopted by the shifting cultivators in the Northeastern region across different states.

In the questionnaire, certain barriers as identified from the review of literature were included for obtaining inputs from the respondents. A total of 11 such barriers were included such as (i) high cost, (ii) method not suitable, (iii) local people's opinion not considered, (iv) adaptation process, not on time, (v) lack of planning, (vi) lack of idea/knowledge, (vii) time limitation, (viii) lack of extension services, (ix) lack of information on climate change, (x) perception of adaptation's ineffectiveness and (xi) economic consequences. The analysis of the responses across the region is reported in Table 5.17.

As per the data from Table 5.17, 82.2% across the region perceived lack of information on climate change followed by 73.8% for economic constraints and 70.8% for lack of extension services as a hindrance to the adaptation of climate change. For other factors, it ranges between 40% and 60% of respondents for the region as a whole. However, there are wide variations in responses across different states in the Northeastern region (refer to Table 5.17). One serious concern raised by the respondents is they believe their opinions are not considered for adaptation policies to climate change. Since every locality has its own rituals, habits and customs; so the opinion of locals regarding such an issue is very important to consider, but 57% of sample households reported that one obstacle to adaptation is ignoring the voice of the local people.

Table 5.17 Perceived barriers to adaptation of climate change in the sample households in the Northeastern region (in percentage)

States	High cost	Method not suitable	Local people opinion not considered	Adaptation process not on time	Lack of planning	Lack of idea/ knowledge	Time limitation	Lack of extension services	Lack of information on climate change	Perception of adaptation ineffectiveness	Economic constraints
1	2	3	4	5	6	7	8	9	10	11	12
Mizoram	10.0	12.0	18.0	17.0	71.0	65.0	70.0	72.0	14.0	16.0	98.0
Tripura	75.6	75.6	76.3	70.8	68.3	73.1	75.6	75.6	76.3	70.8	68.3
Manipur	10.3	29.0	82.0	67.0	12.0	87.0	75.0	59.0	92.0	76.0	93.0
Meghalaya	9.0	80.0	43.0	23.0	67.0	78.0	12.0	49.0	75.0	56.0	90.0
Assam	59.0	12.0	78.0	45.0	76.0	58.0	23.0	87.0	89.0	45.0	98.0
Nagaland	39.0	2.0	12.0	21.0	3.0	5.0	59.9	69.8	74.8	30.8	56.5
Arunachal Pradesh	5.0	50.0	50.0	32.0	54.0	87.5	54.0	57.0	74.0	25.0	12.5
Average NER	41.3	45.8	57.0	48.0	50.2	64.8	56.8	70.8	82.2	57.2	73.8

Source Field Survey, 2019–20

Chapter 6
Trend and Pattern of Weather Variables in Northeastern States of India

6.1 Climatic Variables in the Context of Northeast India

Climate change is known as large variations in climate averages that exist for decades or even for more extended periods. Although the climatic condition has changed all over the world, its impact is not uniform across regions (Trajakovich & Kolkovic, 2009). Since the rate of increase in temperature and variability in precipitation is significantly higher in the Himalayan region, therefore, the potential impacts of climate change on ecology, economy, and society are enormous in this region (Mathison et al., 2013). In recent years, various studies have been done for detecting possible climate trends and changes in India. These studies broadly concluded that during last century there is observed an increasing trend in surface temperature in India (Arora et al., 2005; Dash et al., 2007; Hingane et al., 1985; Kumar et al., 1994; Srivastava et al., 1992). The experts have found that the pattern and magnitude of warming in India has been broadly consistent with the global direction and magnitude over time. The Diurnal Temperature Range (DTR) is the difference between the daily maximum and minimum temperature trends. In India the DTR are reasonably different from those observed in other parts of the world because of the comparatively large increase in 'maximum daily temperature' over a significant portion of India (Kumar et al., 1994; Srivastava et al., 1992).

Similarly, while there is definitive evidence that the trend of rainfall is decreasing in most parts of the world, several studies on the evolving pattern of rainfall over India find no consistent trend of increase or decrease in average rainfall (Kumar et al., 2010; Mooley & Parthasarathy, 1984; Thapliyal & Kulshrestha, 1991). Despite the fact that these studies have contributed significantly to the understanding of the pattern of climatic variables in India, they have concentrated on the study of climatic data for a single station or a group of stations. State-wise analysis on the trend of climatic variables, particularly of rainfall is lacking in Northeast India. Secondly, whatever limited studies have been done those too did not control for the presence

N. Roy et al., *Climate Change, Shifting Cultivation and Livelihood Vulnerabilities in India*, https://doi.org/10.1007/978-3-031-54927-4_6

of serial correlation in long time-series data. The present study aims to analyse the trends in the monthly, annual, and seasonal rainfall in Northeastern Region.

The eight states viz; Assam, Arunachal Pradesh, Manipur, Meghalaya, Mizoram, Nagaland, Sikkim, and Tripura of the NER is spread over 255 168 km^2, which is about 3.85% of the country's area and extends from 21,050′ to 29,034′ Northern latitude, and 85,034′–97,050′ Eastern longitude. From a biogeographical point of view, this region can be divided into the Eastern Himalayas in the North, the Northeast hills in the South, and the Brahmaputra river basin in between (Mani, 1974). The region is characterised by high forest cover, rich biodiversity, fragile mountain ecosystems, high seismicity, and heavy precipitation. The region has a diverse climate ranging from hot and moist tropical in plains and foothills to cold and dry alpine of higher mountains due to influence of locations, topography and altitude. The economy of the Northeastern region is mainly agrarian in nature, with over 70% of the population directly or indirectly dependent on agriculture for livelihood. Like other parts of India, most of the crop areas in the region is under rainfed agriculture, and so is in areas highly vulnerable to climate variability and climate change.

The important weather variables for climate change consist of temperature, humidity, precipitation; air pressure, wind speed, and wind direction that helps to predict the weather. However, in the present study rainfall and temperature have been analysed as the data for these variables are available for all the Northeastern states.

6.2 Methods and Database

In the present study a nonparametric method known as Sen's estimator has been used (Sen, 1968) to estimate the magnitude of trend in rainfall and temperature, and the statistical significance of the trend in the time series was calculated using Mann–Kendall analysis (MK) test (Kendall, 1975; Mann, 1945).

Sen's approach assumes a linear time series pattern and is commonly used to assess the magnitude of trend in time series data of climatic variables such as rainfall, temperature, and so on (Lettenmaier et al., 1994; Partal & Kahya, 2006; Yue & Hashino, 2003). In this method, the slopes (T_i) of all data pairs are first calculated by

$$T_i = \frac{x_j - x_k}{j - k} \quad for\, i = 1, 2, \ldots, N \tag{6.1}$$

where x_j and x_k are data values at time j and k ($j > k$), respectively. The median of these N values of T_i is Sen's estimator of slope, which is calculated as follows:

$$\beta = \begin{cases} T_{\frac{N+1}{2}} & N\, is\, odd \\ \frac{1}{2}\left(T_{\frac{N}{2}} + T_{\frac{N+2}{2}}\right) & N\, is\, even \end{cases} \tag{6.2}$$

A positive value of β indicates an upward (increasing) trend and a negative value indicates a downward (decreasing) trend in the time series.

To ascertain if the Sen's estimator of slope is statistically, nonparametric Mann–Kendall (MK) test has been employed. The MK test is a rank-based non-parametric test that compares the null hypothesis of no trend to the alternative hypothesis of a rising or decreasing trend. Test statistic S is defined as follows.

$$S = \sum_{i=2}^{n} \sum_{j=1}^{t-1} sign(x_i - x_j)$$

where n is the length of the data series x_i and x_j are the sequential data in the series and

$$sign(x_i - x_j) = \begin{cases} -1 \; for \; (x_i - x_j) < 0 \\ 0 \quad for \; (x_i - x_j) = 0 \\ 1 \quad for \; (x_i - x_j) > 0 \end{cases}$$

$$E[S] = 0$$

$$Var(S) = \frac{n(n-1)(2n+5) - \sum_{p-1}^{q} t_p(t_p - 1)(2t_p + 5)}{18}$$

The number of ties for the pth value is t_p, and the number of tied values is q. The second term in the variance formula is for tied censored data. Standardised test statistics Z is computed by

$$z = \begin{cases} \frac{s-1}{\sqrt{Var(s)}} \\ 0 \\ \frac{s+1}{\sqrt{Var(s)}} \end{cases}$$

If the computed value of $|Z| > z_{\alpha/2}$, the null hypothesis (H_0) is rejected at α level of significance in a two-sided test. In this analysis, the null hypothesis was tested at 95% confidence level. Modified Mann–Kendall Test for serially correlated data following Yue and Wang (2004) Variance Correction Approach has been used.

The collected state-wise monthly/seasonal data on rainfall and temperature from the Indian Meteorological Department (IMD) for all the eight states of NER for the period 1901–2017 has been used in this study for estimation. The Indian Meteorological Department (IMD) is an agency of the Ministry of Earth Sciences of the Government of India which solders the responsibility for meteorological observations, weather forecasting and seismology in India. IMD is headquartered in Delhi and operates hundreds of observation stations across India and Antarctica. Meticulous quality checks and pre-processing are done by IMD before releasing data so as to provide error-free data for analysis and design. Therefore, IMD data is regarded as the most reliable long series of data for the study area. Some of the data have also been collected from IndiaStat.com publications based on IMD sources.

6.3 Trend Analysis of Monthly Rainfall Data

As expected, rainfall trends show large variability in magnitude and direction of trend from one state to another. We have estimated 96 values (12 months × eight states) of Sen's coefficient for all eight states of NER. The analysis reveals that 61 values of Sen's coefficients were negative, 43 values were positive, and one value was zero (refer Table 6.1). Out these 61 negative values, 42 values were statistically significant at least at 10% level of significance. On the other hand, 25 of the positive values were found to be statistically significant. This clearly indicates that the decreasing trend of rainfall is dominating the picture. Nagaland, Manipur and Assam are the three states for which the trend of rainfall is negative for most of the months. Conversely, Arunachal Pradesh and Meghalaya have experienced an increasing trend of rainfall for most of the months.

The actual and normal rainfall during 2015–2018 across different states is reported in Table 6.1 for a glimpse on the situation. The state wise rainfall distribution across districts during 2017–2018 shows that in 70.8% of the districts in the region there was no deficient in rainfall in 2017 which increased to 82.8%. This is a encouraging situation considering the fragility of the region in environmental degradation. In 25% of the districts there were large deficient in rainfall in 2017 and increased to 28.6% in 2018. The detailed account is reported in Table 6.2.

In respect of the availability of inland water resources are concerned it is observed that in the Northeastern region the highest amount i.e. 32.5% of total inland water resources consisted of beels followed by 29.5% belonging to tanks and ponds and 26.1% consisted of derelict water. At all India level highest proportion i.e. 41.8%

Table 6.1 Sub divisional actual and normal rainfall in North Eastern States

Meteorological sub-divisions	2015		2016		2017		2018	
	Actual	Normal	Actual	Normal	Actual	Normal	Actual	Normal
1	2	3	4	5	6	7	8	9
Arunachal Pradesh	2745.0	2933.7	2706.9	2933.7	2745.3	2933.7	2032.5	2933.7
Assam and Meghalaya	2509.4	2624.9	2266.9	2624.9	2711.7	2624.9	1903.1	2624.9
Nagaland, Mizoram, Manipur and Tripura	1947.4	2278.0	1956.4	2278.0	1165.8	1296.3	960.0	1296.3
Sub-Himalayan West Bengal and Sikkim	2501.7	2708.9	2624.8	2708.9	1112.0	1205.6	860.6	1205.6
NER average	2425.9	2636.4	2388.8	2636.4	1933.7	2015.1	1439.1	2015.1

Source India Meteorological Department, Ministry of Earth Sciences as published in as compiled from EnviStats India 2020, Vol. I, Social Statistics Division, Ministry of Statistics and Programme Implementation, Govt. of India

Table 6.2 State-wise rainfall distribution of districts in Northeastern Region, 2017 and 2018

State	LE		E		N		D		LD		ND		Total		% of normal rainfall
	2017	2018	2017	2018	2017	2018	2017	2018	2017	2018	2017	2018	2017	2018	
1	2	3	4	5	6	7	8	9	10	11	12	13	14	15	16
Arunachal Pradesh	1	2	0	0	3	5	7	5	2	1	3	3	16	16	19
Assam	0	0	2	3	11	18	12	5	0	0	2	1	27	27	41
Manipur	0	1	1	0	0	1	2	2	1	0	5	5	9	9	–
Meghalaya	0	1	0	0	3	1	3	2	0	0	1	3	7	7	43
Mizoram	0	1	2	1	1	1	1	0	0	0	5	6	9	9	11
Nagaland	0	2	2	0	1	1	7	1	0	1	1	6	11	11	9
Sikkim	0	0	1	1	3	3	0	0	0	0	0	0	4	4	75
Tripura	0	0	0	2	2	2	2	0	0	0	0	0	4	4	50
NE total	**1 (8.3)**	**7 (28.0)**	**8 (8.0)**	**7 (9,0.0)**	**24 (8.3)**	**32 (10.0)**	**34 (13.9)**	**15 (7.4)**	**3 (25.0)**	**2 (28.6)**	**17 (70.8)**	**24 (82.8)**	**87 (12.8)**	**87 (13.2)**	35.43

Source India Meteorological Department, M/o Earth Sciences as mentioned in EnviStats India 2020, Vol. I, Social Statistics Division, Ministry of Statistics and Programme Implementation, Govt. of India. *Note* The rainfall categories have been modified w.e.f. 2016

Category Percentage deviation from Long Period Average, *LE* large excess + 60% or more, *E* excess + 20% to + 59%, *D* deficient – 20% to – 59%, *LD* large deficient – 60% to – 90%, *N* normal + 19% to – 19%. *ND* no deficient

consisted of reservoirs followed by 29.4% belongs to tanks and ponds and 13.7% consisted of brackish water. Due to the non-availability of reservours the irrigation facilities in the Northeastern states could not be expanded in truest spirit. The analysis is reported in Tables 6.3 and 6.4.

6.3.1 Trend Analysis of Seasonal Rainfall

In this analysis each year was divided into four climatic seasons (Rao, 1981) to see if there were any patterns in the data at the seasonal scale. These were winter (January–February), pre-monsoon (March–May), southwest monsoon (June–October), and post-monsoon (November–December). It is important to note here that the southwest monsoon season in this region lasts for five months (June–October) in contrast to 4 months (June–September) for the whole of India, while the post-monsoon season lasts for only two months. In the present study, the trend analysis of seasonal rainfall is analysed by drawing scatter plots of seasonal rainfall data (see Charts 6.1, 6.2, 6.3 and 6.4).

It is observed that apart from the pre-monsoon season, the trend of rainfall is roughly negative in Northeastern region of India in all other seasons. In pre-monsoon season, the trend line is almost parallel to the horizontal axis, which indicates that we do not have any noticeable trend of rainfall in pre-monsoon season. Speaking about the magnitude of rainfall, southwest monsoon is observed to have the highest magnitude of rainfall followed by pre-monsoon and post-monsoon seasons. Finally, year to year variation in rainfall is observed to be the highest in post-monsoon season marginally followed by the winter season. Conversely, the year wise variation in rainfall is observed to be the lowest in southwest season.

In order to analyse the trend of seasonal rainfall across different states, Sen's slope has been estimated in this study. The estimated results are reported in Table 6.5.

It is observed that, for Arunachal Pradesh (AP), the estimated slopes are all positive and statistically significant. This indicates that the trend of rainfall in AP has increased in all seasons over the study period. However, the magnitude of Sen's coefficient is not uniform across seasons: the value of the coefficient is estimated to be highest in southwest monsoon and lowest in post-monsoon seasons. The picture of Sen's coefficient is just opposite in Assam: here all the values are positive and are statistically significant baring the case of the post-monsoon season. This reveals that the trend of rainfall has decreased in Assam: the southwest monsoon season being the most suffering season of the lot. Similarly, the Sen's coefficients are estimated to be positive in all seasons for Manipur. Therefore, Manipur also shares the same fate that of Assam: the falling trend of rainfall in all seasons. Fourthly, the trend of rainfall has exhibited a decreasing trend in winter and increasing trend in post-monsoon seasons for Meghalaya. The trend of rainfall in pre-monsoon and southwest monsoon seasons is found to be statistically insignificant for the state. Fifthly, Mizoram and Nagaland have exhibited a decreasing trend of rainfall in all seasons which are statistically significant. Like Assam, these states also experienced the highest fall in rainfall

Table 6.3 State-wise inland water resources

States	Rivers and canals (length in km)	Reservoirs (ha)	Tanks and ponds (ha)	Brackish water (ha)	Beels (ha)	Oxbow lakes (ha)	Derelict water (ha)	Other than rivers and canals (ha)	Total (ha)
1	2	3	4	5	6	7	8	9	10
Arunachal Pradesh	30.0	160.0	3625.0	0	0	5.0	11,864.0	0	15,654.0
Assam	4820.0	2000.0	73,065.0	0	100,815.0	0	86,204.0	0	262,084.0
Manipur	3360.0	2142.0	11,442.0	0	24,333.0	0	4728.0	0	42,745.0
Meghalaya	3904.12	8000.0	2000.0	0	220.93	61.45	53.94	0	10,336.32
Mizoram	1100.0	8100.0	5468.34	0	0	0	0	0	13,568.34
Nagaland	1600.0	2258.0	3425.5	0	0	1700.0	0	0	7383.5
Sikkim	900.0	850.0	15.7	0	3000.0	0	0	0	3865.7
Tripura	1266.71	5000.0	17,552.07	0	0	0	361.44	0	22,913.51
NE total	16,980.83 (4.3)	28,510 (7.2)	116,593.6 (29.5)	0 (0)	128,368.9 (32.5)	1766.45 (0.4)	103,211.4 (26.1)	0 (0)	395,431.2 (100.0)
All India	201,495.7 (2.4)	3,524,724.2 (41.8)	2,478,262.2 (29.4)	1,160,162.5 (13.7)	424,850.9 (5.0)	117,800.5 (1.4)	230,136.4 (2.7)	300,724.5 (3.6)	8,438,156.9 (100.0)

Source Handbook of Fisheries Statistics, 2018, Ministry of Fisheries, Animal Husbandry and Dairying, as compiled from EnviStats India 2020, Vol. I, Social Statistics Division, Ministry of Statistics and Programme Implementation, Govt. of India

Table 6.4 Sen's estimator of slope (mm/year) for monthly rainfall

	AP	AS	MAN	MEGH	MIZ	NAG	TRI	SK
1	2	3	4	5	6	7	8	9
Jan	0.02	– 0.01	– 0.04	0.00	– 0.02	– 0.04	0.04	– 0.01
Feb	0.09	– 0.04	– 0.13	– 0.05	– 0.09	– 0.11	– 0.02	– 0.04
Mar	0.23	– 0.02	– 0.03	0.06	0.18	– 0.10	0.27	0.12
Apr	0.36	– 0.01	– 0.57	0.31	0.06	– 0.33	0.61	0.14
May	0.28	– 0.41	– 1.24	– 0.38	– 0.92	– 0.65	0.53	– 0.15
Jun	– 0.14	– 1.47	– 3.28	– 0.59	– 2.04	– 1.74	– 1.71	– 1.31
Jul	0.54	– 0.19	– 0.88	1.55	– 0.39	– 0.37	– 2.47	– 0.09
Aug	– 0.05	– 0.70	– 1.38	0.15	– 0.69	– 0.64	– 1.37	– 0.76
Sep	0.61	0.02	– 0.41	0.62	0.31	– 0.06	– 0.50	0.10
Oct	0.12	– 0.28	– 0.70	0.17	– 0.30	– 0.30	– 0.26	– 0.22
Nov	0.00	– 0.04	– 0.13	0.01	– 0.09	– 0.10	0.00	– 0.08
Dec	0.02	0.02	0.01	0.01	– 0.01	0.02	0.00	0.00

Note (1) Bold values indicate statistical significance at least at 90% confidence level as per the Mann–Kendall test (+ for increasing trend and – for decreasing trend). (2) *AP* Arunachal Pradesh, *AS* Assamn, *MAN* Manipur, *MEGH* Meghalaya, *MIZ* Mizoram, *NAG* Nagaland, *SK* Sikkim, and *TR* Tripura

Source Computed by authors from IMD data

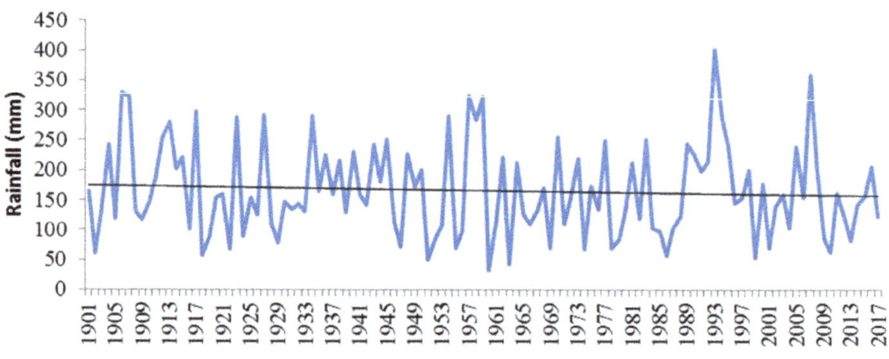

Chart 6.1 Trend of rainfall in Northeast: winter. *Source* Compiled from IMD data

in the southwest monsoon and lowest fall in rainfall in post-monsoon seasons. In Sikkim, the Sen's coefficient is found to be significant only for pre-monsoon and southwest monsoon seasons. In the pre-monsoon seasons, the positive value of the coefficient indicates that the trend of rainfall has an increasing trend for the season, whereas the negative value of the coefficient in southwest monsoon reveals that the trend of rainfall has decreased in this season. Finally, Tripura also has exhibited a decreasing trend of rainfall in all but pre-monsoon season. The Sen's coefficient of rainfall is found to be statistically insignificant in the pre-monsoon season.

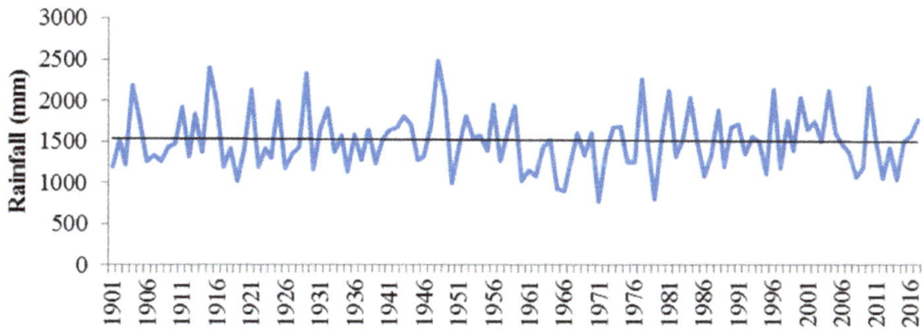

Chart 6.2 Trend of rainfall in Northeast: pre-monsoon. *Source* Compiled from IMD data

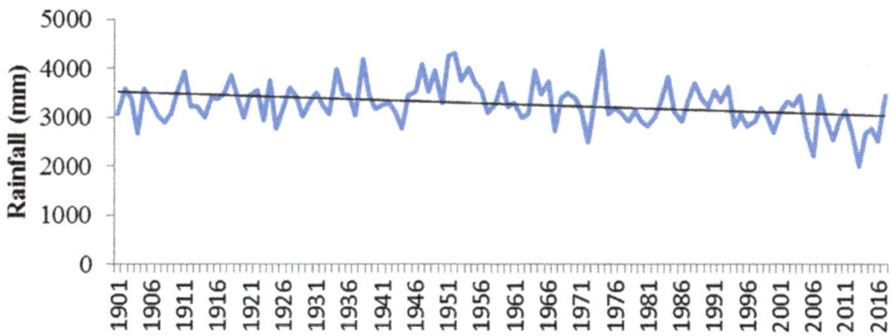

Chart 6.3 Trend of rainfall in Northeast: southwest monsoon. *Source* Compiled from IMD data

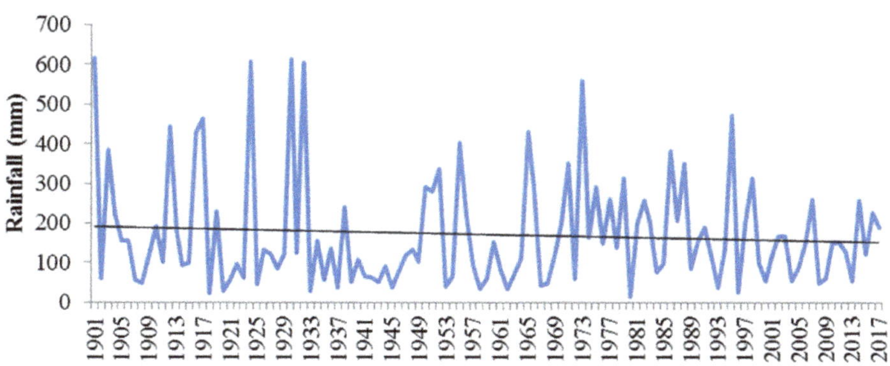

Chart 6.4 Trend of rainfall in Northeast: post-monsoon. *Source* Compiled from IMD data

Table 6.5 Sen's estimator of slope (mm/year) for seasonal rainfall

	Winter	Pre-monsoon	Southwest monsoon	Post-monsoon
1	2	3	4	5
AP	**0.11**	**0.85**	**1.04**	**0.03**
AS	**– 0.07**	**– 0.38**	**– 2.51**	– 0.04
MAN	**– 0.21**	**– 1.76**	**– 5.30**	**– 0.10**
MEGH	**– 0.06**	0.07	1.41	**0.04**
MIZ	**– 0.10**	**– 0.49**	**– 3.07**	**– 0.08**
NAG	**– 0.15**	**– 1.04**	**– 3.14**	**– 0.09**
SK	0.02	**1.44**	**– 6.49**	0.05
TR	**– 0.06**	0.31	**– 2.49**	**– 0.06**

Bold values indicate statistical significance at least at 90% confidence level as per the Mann–Kendall test (+ for increasing trend and – for decreasing trend)
AP Arunachal Pradesh, *AS* Assamn, *MAN* Manipur, *MEGH* Meghalaya, *MIZ* Mizoram, *NAG* Nagaland, *SK* Sikkim, and *TR* Tripura

6.3.2 *Trend Analysis of Annual Rainfall*

This section begins with a comparative analysis of annual rainfall data of India and NER. To analyse this comparison, calculation of annual average rainfall data of India and NER from 1901 to 2013 has been made. The calculated data have been plotted in a trend line figure to make the comparison visually convenient (refer Chart 6.5).

It is evident from Chart 6.5 that the annual average rainfall is significantly higher for NER than all India average during the entire study period. Secondly, the trend of annual average rainfall has remained almost stable for India while that of NER has shown a gradual decline since 1960. The Sen's coefficient estimated for annual data of NER states is reported in Table 6.5.

The analysis from the values in Table 6.6 shows that the coefficient of Sen's estimator is negative and statistically significant for Assam, Manipur, Mizoram, Nagaland, Tripura, and Sikkim. This indicates that six of the NER states have experienced a decreasing trend of rainfall during the study period. Out of these six states, the

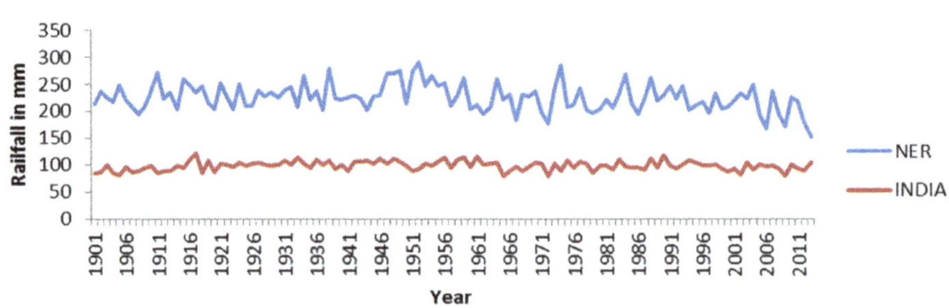

Chart 6.5 Trend of annual average rainfall (mm). *Source* Compiled from IMD data

Table 6.6 Sen's estimator of slope (mm/year) for annual rainfall

States	Sen's coefficient
1	2
AP	**1.67**
AS	**– 2.78**
MAN	**– 6.51**
MEGH	1.20
MIZ	**– 3.77**
NAG	**– 3.93**
TR	**– 5.12**
SK	**– 2.37**

Bold values indicate statistical significance at least at 90% confidence level as per the Mann–Kendall test (+ for increasing trend and – for decreasing trend)
Source Estimated by authors

negative value of Sen's coefficient is estimated to highest for Manipur marginally followed by Tripura. Therefore, the decreasing trend of rainfall is more in these two states. The two other states where the coefficient turns out to be positive are Arunachal Pradesh and Meghalaya. But the Sen's coefficient for Meghalaya is not statistically significant, indicating that there no significant trend of annual rainfall for the state of Meghalaya. The annual rainfall has remained almost stable during the study period. Arunachal Pradesh is the only state where the Sen's coefficient is positive and significant. This implies that the annual trend of rainfall is on the rise in Arunachal Pradesh during the study period.

6.4 Trend Analysis of Temperature

In this section estimation of the trend in temperature variable in all the Northeastern states of India over the period 1901–2017 is reported. The data for temperature have been obtained from Indian Meteorological Department (IMD) for all the eight states of NER for the period 1901–2017.

From the analysis of the estimate presented in Table 6.7, it is found variations in temperature throughout the year while it is highest during the winter season and lowest during the southwest monsoon in the state of Assam. The estimated slopes are all positive and statistically significant. This indicates that the trend of temperature in the state of Assam has increased in all seasons over the study period. However, the magnitude of Sen's coefficient is not uniform across seasons: the value of the coefficient is estimated to be highest in winter followed by pre-monsoon, southwest monsoon and post-monsoon. Further, the variability is more for maximum temperature achieved compared to the minimum in the temperature records.

Table 6.7 Sen's estimator of slope for temperature-case of Assam

Seasons	Max	Min	Sen's slope	Z-value
1	2	3	4	5
Annual	27.72 (0.41)	18.26 (0.39)	0.0078	15.10*
Winter	23.32 (0.74)	11.34 (0.67)	0.0096	14.48*
Pre-monsoon	28.93 (0.70)	18.42 (0.55)	0.0078	13.73*
Southwest monsoon	29.72 (0.33)	22.85 (0.31)	0.0043	9.40*
Post monsoon	24.73 (0.66)	13.56 (0.67)	0.009	11.68*

Note (1) *values indicate statistical significance at least at 90% confidence level as per the Mann–Kendall test. (2) Values within parenthesis indicate standard deviation. *Source* Estimated by authors

As evident from the estimate in Table 6.8, the temperature has been increasing since 1901 in the state of Arunachal Pradesh in all the seasons of the year. The significant z-values validating the positive trend results signify that without a doubt, there has been a rise in temperature in the state. The slope coefficient is estimated to be highest for the winter season while it is lowest for the southwest monsoon. There is also wide variation in temperature in the state ranging from an average minimum of 5.960 to a maximum of 17.310 in winter while from 18.630 to 26.180 during the southwest monsoon. The annual temperature was estimated to vary between averages of 11.540–21.530 in Arunachal Pradesh.

Similar to Assam and Arunachal Pradesh, the trend in temperature in Manipur is also positive throughout the study period as is observed from Table 6.9. With a 90% confidence, we have estimated highest slope for the winter season and lowest for post-monsoon. The lowest temperature in the state hovers around 8°–9° while

Table 6.8 Sen's estimator of slope for temperature-case of Arunachal Pradesh

Seasons	Max	Min	Sen's slope	Z-value
1	2	3	4	5
Annual	21.53	11.54	0.007	16.31*
Winter	17.31	5.96	0.009	15.33*
Pre-monsoon	23.16	13.41	0.008	17.08*
Southwest monsoon	26.18	18.63	0.004	11.43*
Post monsoon	19.46	8.16	0.007	10.99*

Notes (1) According to the Mann–Kendall test, *values suggest statistical significance at least at a 90% confidence level. (2) Values within parenthesis indicate standard deviation. *Source* Estimated by authors

the highest around 26° on an average. Thus, there have been substantial fluctuations in temperature across seasons.

Unlike the earlier three states, the Mann–Kendall test provided the highest trend in post-monsoon temperature for the state of Mizoram. The estimated values are presented in Table 6.10. The overall annual trend has been found positive, with the lowest trend value for pre-monsoon. The lowest temperature in the state was estimated to be around 12° on average, while the highest to be around 31.

The estimated result for the state of Tripura is depicted in Table 6.11. The estimate shows a rise in temperature over the years and the rise is more prominent in winter compared to other seasons of the year. Looking at the maximum and minimum figures, we found the minimum temperature during winter to be around 13° and the highest temperature around 26°. The hottest months, on the other hand, in the state, seem to be those of pre-monsoon ones.

The lowest temperature in the state of Nagaland has been found to be 7.690 during winter while the highest (26.160) in southwest monsoon indicating oscillations in temperature as evident from the estimated values depicted in Table 6.12. The slope coefficient is found to be highest for winter and post-monsoon followed

Table 6.9 Sen's estimator of slope for temperature-case of Manipur

Seasons	Max	Min	Sen's slope	Z-value
1	2	3	4	5
Annual	23.90	13.89	0.006	13.45*
Winter	21.12	8.52	0.006	10.43*
Pre-monsoon	26.22	15.87	0.004	11.14*
Southwest monsoon	26.07	19.55	0.004	8.36*
Post monsoon	22.18	11.61	0.001	11.26*

Notes (1) According to the Mann–Kendall test, *values suggest statistical significance at least at a 90% confidence level. (2) Values within parenthesis indicate standard deviation. *Source* Estimated by authors

Table 6.10 Sen's estimator of slope for temperature-case of Mizoram

Seasons	Max	Min	Sen's slope	Z-value
1	2	3	4	5
Annual	28.15	18.58	0.005	11.96*
Winter	25.80	13.41	0.005	10.82*
Pre-monsoon	30.90	20.93	0.003	8.84*
Southwest monsoon	29.55	19.38	0.006	7.00*
Post monsoon	26.37	20.58	0.008	12.13*

Notes (1) According to the Mann–Kendall test, *values suggest statistical significance at least at a 90% confidence level. (2) Values within parenthesis indicate standard deviation. *Source* Estimated by authors

Table 6.11 Sen's etimator of slope for temperature-case of Tripura

Seasons	Max	Min	Sen's slope	Z-value
1	2	3	4	5
Annual	29.18	19.30		
Winter	26.57	13.60	0.0145	4.38*
Pre-monsoon	32.01	22.19	0.0098	5.02*
Southwest monsoon	30.77	24.90	0.0064	4.96*
Post monsoon	27.40	16.51	0.0105	4.66*

Notes (1) According to the Mann–Kendall test, *values suggest statistical significance at least at a 90% confidence level. (2) Values within parenthesis indicate standard deviation. *Source* Estimated by authors

by pre-monsoon and southwest monsoon. The comparatively maximum temperature recorded during southwest monsoon seasons.

Annual trend in temperature in the case of the state of Meghalaya is positive similar to all other Northeastern States. This is evident from the data as depicted in Table 6.13. Highest trend value is found for the post-monsoon season which is tailed by winter, pre-monsoon and southwest monsoon. Although there are fluctuations in temperature throughout different seasons, the variability seemed to be lowest in southwest monsoon with a minimum temperature of 23.76° and a maximum of 29.93°.

For the state of Sikkim, Sen's estimators of slope for temperature reveal an increasing trend for annual as well as all the four seasons of the year. The trend is appraised to be highest in winter and post-monsoon and lowest in the southwest monsoon. The average lowest temperature in the state was recorded at 5.3 in winter while southwest monsoon had the highest record at 26.09°. This is reported in Table 6.14.

Thus, an increasing trend in temperature for all the Northeastern states of India has been observed during the period under study. The coldest states are Sikkim,

Table 6.12 Sen's estimator of slope for temperature-case of Nagaland

Seasons	Max	Min	Sen's slope	Z-value
1	2	3	4	5
Annual	23.10	12.94	0.007	10.98*
Winter	19.92	7.69	0.008	11.53*
Pre-monsoon	24.89	14.59	0.006	14.12*
Southwest monsoon	26.16	19.12	0.005	10.21*
Post monsoon	21.47	10.36	0.008	9.12*

Notes (1) According to the Mann–Kendall test, *values suggest statistical significance at least at a 90% confidence level. (2) Values within parenthesis indicate standard deviation. *Source* Estimated by authors

Table 6.13 Sen's estimator of slope for temperature-case of Meghalaya

Seasons	Max	Min	Sen's slope	Z-value
1	2	3	4	5
Annual	27.68	17.51	0.006	12.65*
Winter	24.59	11.71	0.008	13.61*
Pre-monsoon	30.42	19.84	0.004	6.89*
Southwest monsoon	29.93	23.76	0.002	6.68*
Post monsoon	25.78	14.72	0.010	11.05*

Notes (1) According to the Mann–Kendall test, *values suggest statistical significance at least at a 90% confidence level. (2) Values within parenthesis indicate standard deviation. *Source* Estimated by authors

Table 6.14 Sen's estimator of slope for temperature-case of Sikkim

Seasons	Max	Min	Sen's slope	Z-value
1	2	3	4	5
Annual	22.35	11.35	0.007	15.90*
Winter	17.57	5.30	0.009	19.75*
Pre-monsoon	24.92	13.62	0.008	9.68*
Southwest monsoon	26.09	18.17	0.002	5.98*
Post monsoon	20.85	8.25	0.009	12.24*

Notes (1) According to the Mann–Kendall test, *values suggest statistical significance at least at a 90% confidence level. (2) Values within parenthesis indicate standard deviation. *Source* Estimated by authors

Arunachal Pradesh, Nagaland and Manipur respectively in view of the lowest recorded temperature. Tripura, Assam and Mizoram, on the other hand, recorded the average highest temperatures during pre-monsoon and southwest monsoon. Finally, the region recorded a wide variation in temperature from a lowest of 5 to highest of 32°. However, it is also to be noticed that these figures are average for 1901–2017 and thus only reveal approximations for the seasons. But in reality, there are some months with heavy snowfall in places of Sikkim and Arunachal Pradesh and the temperature reaching to 0° while in some other places of the region temperature in some days reach to 40°.

References

Arora, M., Goel, N. K., & Singh, P. (2005). Evaluation of temperature trends over India/Evaluation de tendances de température en Inde. *Hydrological Sciences Journal, 50*(1).

Dash, S. K., Jenamani, R. K., Kalsi, S. R., & Panda, S. K. (2007). Some evidence of climate change in twentieth-century India. *Climatic Change, 85*(3–4), 299–321.

Hingane, L. S., Kumar, K. R., & Murty, B. V. R. (1985). Long-term trends of surface air temperature in India. *Journal of Climatology, 5*(5), 521–528.

Kendall, M. G. (1975). *Rank correlation methods*. Charles Griffin.

Kumar, K. R., Kumar, K. K., & Pant, G. B. (1994). Diurnal asymmetry of surface temperature trends over India. *Geophysical Research Letters, 21*(8), 677–680.

Kumar, V., Jain, S. K., & Singh, Y. (2010). Analysis of long-term rainfall trends in India. *Hydrological Sciences Journal-Journal Des Sciences Hydrologiques, 55*(4), 484–496.

Lettenmaier, D. P., Wood, E. F., & Wallis, J. R. (1994). Hydro-climatological trends in the continental United States, 1948–88. *Journal of Climate, 7*(4), 586–607.

Mani, M. S. (1974). Biogeographical evolution in India. In *Ecology and biogeography in India* (pp. 698–724). Springer.

Mann, H. B. (1945). Nonparametric tests against trend. *Econometrica: Journal of the Econometric Society*, 245–259.

Mathison, C., Wiltshire, A., Dimri, A. P., Falloon, P., Jacob, D., Kumar, P., Jacob, D., Kumar, P., Moors, E., Ridley, J., Siderius, C., Stoffel, M., & Yasunari, T. (2013). Regional projections of North Indian climate for adaptation studies. *Science of the Total Environment, 468*, S4–S17.

Mooley, D. A., & Parthasarathy, B. (1984). Fluctuations in all-India summer monsoon rainfall during 1871–1978. *Climatic Change, 6*(3), 287–301.

Partal, T., & Kahya, E. (2006). Trend analysis in Turkish precipitation data. *Hydrological Processes, 20*(9), 2011–2026.

Rao, Y. P. (1981). The climate of the Indian subcontinent. In K. Takahasi & H. Arakawa (Eds.), *World survey of climatology* (Vol. 9, pp. 67–182). Elsevier.

Sen, P. K. (1968). Estimates of the regression coefficient based on Kendall's tau. *Journal of the American Statistical Association, 63*(324), 1379–1389.

Srivastava, H. N., Dewan, B. N., Dikshit, S. K., Rao, G. P., Singh, S. S., & Rao, K. R. (1992). Decadal trends in climate over India. *Mausam, 43*(1), 7–20.

Thapliyal, V., & Kulshrestha, S. M. (1991). Climate changes and trends over India. *Mausam, 42*(4), 333–338.

Trajkovic, S., & Kolakovic, S. (2009). Evaluation of reference evapotranspiration equations under humid conditions. *Water Resources Management, 23*(14), 3057.

Yue, S., & Hashino, M. (2003). Long term trends of annual and monthly precipitation in Japan 1. *JAWRA Journal of the American Water Resources Association, 39*(3), 587–596.

Yue, S., & Wang, C. (2004). The mann-kendall test modified by effective sample size to detect trend in serially correlated hydrological series. *Water Resources Management 18*, 201–218. https://doi.org/10.1023/B:WARM.0000043140.61082.60.

Chapter 7
Household Level Livelihood Vulnerabilities in Northeastern States of India

7.1 Livelihood Vulnerability Index (LVI): Composite Index Approach

The LVI was initially intended to give policymakers, planners and development agencies with a scientific instrument to comprehend the role of physical, populations and social factors to climate change vulnerability. This gives a fantastic approach where organisers can refine and centre their investigation to suit the requirements of each topographical area. Besides the overall composite index, separate sectoral vulnerability indices can be prepared to recognize potential areas for intervention (Hahn et al., 2009). It takes into consideration of seven major components: Socio-Demographic Profile, Livelihood Strategies, Social Networks, Health, Food, Water, and Natural Disasters and Climate Variability. Each of the seven major parts of the LVI comprises several indicators or sub-components (refer Table 7.1). These sub-components were developed based on a detailed review of the literature on each major component (Hahn et al., 2009; Mainlay & Tan, 2012; Shah et al., 2013; Panthi et al., 2016), as well as the practicality of collecting the needed data through household surveys. Table 7.1 provides a description about the quantification of each sub-component, the survey question used to collect the data, the original source of the survey question, and potential sources of bias. A brief background of the sample farmers is depicted in Table 7.2.

The quantitative approach used in the present study to calculate the LVI was adapted from Hahn et al. (2009) and is summarised here for expediency. Since each of the sub-components was measured on a different scale; therefore, it was first essential to standardize them for comparability. To do this we used a formula which is developed and compiled by the United Nations to measure human development index. This formula is basically a ratio of the difference of the actual value and the minimum of a variable, and the range of maximum and minimum value of the same variable:

N. Roy et al., *Climate Change, Shifting Cultivation and Livelihood Vulnerabilities in India*, https://doi.org/10.1007/978-3-031-54927-4_7

Table 7.1 Major components and sub-components comprising the livelihood vulnerability index (LVI)

Major components	Sub-components	Explanation of sub-components
Socio-demographic profile	Dependency ratio (DR)	Ratio of the population under 15 and over 65 years of age to the population between 19 and 64 years of age
	Percent of female-headed households (FH)	Percentage of households where the primary adult is female. If a male head is away from the home > 6 months per year the female is counted as the head of the household
	Average age of female head of household (Age)	Age in numbers
	Percent of households where head of household has not attended school (SCH)	Percentage of households where the head of the household reports that they have attended 0 years of school. Percentage of households that have at least 1 orphan living in their home. Orphans are children < 18 years old who have lost one or both parents
Livelihood	Percent of households with family member working in a different community (WDC)	Percentage of households that report at least 1 family member who works outside of the community for their primary work activity
	Percent of households dependent solely on agriculture as a source of income (SA)	Percentage of households that report only agriculture as a source of income
	Average agricultural Livelihood Diversification Index (LDI)	The inverse of (the number of agricultural livelihood activities + 1) reported by a household, e.g., a household that farms, raises animals, and collects natural resources will have a Livelihood Diversification Index = 1/(3 + 1) = 0.25
Health	Average time to health facility (ATH)	Average time it takes the households to get to the nearest health facility
	Percent of households with family member with chronic illness (MCI)	Percentage of households that report at least 1 family member with chronic illness. Chronic illness was defined subjectively by respondent

(continued)

Table 7.1 (continued)

Major components	Sub-components	Explanation of sub-components
	Percent of households where a family member had to miss work or school in the last 2 weeks due to illness (MWS)	Percentage of households that report at least 1 family member who had to miss school of work due to illness in the last 2 weeks
	Average malaria exposure * Prevention index (MEI)	Months reported exposure to malaria * Owning at least one bednet indicator (have bednet = 0.5, no bednet = 1) (e.g., Respondent reported malaria is a problem January–March and they do not own a bednet = 3 * 1 = 3)
Social networks	Average receive: give ratio (RGR)	Ratio of (the number of types of help received by a household in the past month + 1) to (the number of types of help given by a household to someone else in the past month + 1)
	Average borrow: lend money ratio (BLR)	Ratio of a household borrowing money in the past month to a household lending money in the past month, e.g., If a household borrowed money but did not lend money, the ratio = 2:1 or 2 and if they lent money but did not borrow any, the ratio = 1:2 or 0.5
	Percent of households that have not gone to their local government for assistance in the past 12 months (NGA)	Percentage of households that reported that they have not asked their local government for any assistance in the past 12 months
Food	Percent of households dependent solely on family farm for food (AD)	Percentage of households that get their food primarily from their personal farms
	Average number of months households struggle to find food (SF)	Average number of months households struggle to obtain food for their family
	Average Crop Diversity Index (CDI)	The inverse of (the number of crops grown by a household + 1). e.g., A household that grows pumpkin, maize, nhemba beans, and cassava will have a Crop Diversity Index = 1/(4 + 1) = 0.20

(continued)

Table 7.1 (continued)

Major components	Sub-components	Explanation of sub-components
	Percent of households that do not save crops (NSC)	Percentage of households that do not save crops from each harvest
	Percent of households that do not save seeds (NSS)	Percentage of households that do not have seeds from year to year
Water	Percent of households reporting water conflicts (WC)	Percentage of households that report having heard about conflicts over water in their community
	Percent of households that utilize a natural water source (NWS)	Percentage of households that report a creek, river, lake, pool, or hole as their primary water source
	Average time to water source (TWS)	Average time it takes the households to travel to their primary water source
	Percent of households that do not have a consistent water supply (NCW)	Percentage of households that report that water is not available at their primary water source everyday
Natural disasters and climate variability	Average number of flood, drought, and cyclone events in the past 10 years (ND)	Total number of floods, droughts, and cyclones that were reported by households in the past 10 years
	Percent of households with an injury or death as a result of recent natural disasters (ID)	Percentage of households that reported either an injury to or death of one of their family members as a result of the most severe flood, drought, or cyclone in the past 10 years
	Mean standard deviation of monthly average of average maximum daily temperature (years: 2008–2018) (SDXT)	Standard deviation of the average daily maximum temperature by month between 2008 and 2018 was averaged for each province

(continued)

Table 7.1 (continued)

Major components	Sub-components	Explanation of sub-components
	Mean standard deviation of monthly average of average minimum daily temperature (years: 2008–2018) (SDMT)	Standard deviation of the average daily minimum temperature by month between 2008 and 2018 was averaged for each province
	Mean standard deviation of monthly average precipitation (years: 2008–2018) (SDPR)	Standard deviation of the average monthly precipitation between 2008 band 2018 was averaged for each province

Table 7.2 Basic background of the farmers: state-wise analysis (in percentage)

Variable		Arunachal Pradesh	Assam	Manipur	Meghalaya	Mizoram	Nagaland	Tripura
1	2	3	4	5	6	7	8	9
Gender	Male	53.14	49.33	52.99	47.26	50.33	50.97	50.63
	Female	46.86	50.67	47.01	52.74	49.67	49.03	49.37
Age group (in years)	Below-30	57.60	61.00	58.40	67.80	52.10	56.60	56.90
	30–40	13.80	12.30	22.80	11.50	15.40	14.20	17.20
	40–50	14.00	12.30	7.70	10.60	11.80	12.70	12.50
	50 and above	14.70	14.40	11.10	10.10	20.70	16.60	13.40
Level of education	No formal education	0	0	16.5	18.7	2	14.6	21.7
	Primary	11.2	24.5	27.4	20.6	24.8	14.8	27.2
	Secondary	44.4	52.8	33.7	48.7	50.5	54.5	46.1
	College and above	44.4	22.6	22.4	12	22.8	16	5.1
Economic status	Below poverty line	93.58	53.33	95.45	82.1	66.3	55.9	68.9
	Above poverty line	6.42	46.67	4.55	17.9	33.7	44.1	31.1

Source Field Survey, 2018–20

$$Index_{S_d} = \frac{S_d - S_{\min}}{S_{\max} - S_{\min}} \tag{7.1}$$

Here, S_d is the actual value of a variable for district d, and S_{min} and S_{max} are its minimum and maximum values. These minimum and maximum values were used to transform this indicator into a standardized value between 0 and 1 so that it could be integrated with the other components of the LVI. For variables that measure frequencies, such as the 'percent of female-headed households', the minimum value was set at 0 and the maximum at 100%. Some sub-components such as the 'average livestock diversity index' were constructed as the inverse of the crude indicator because it was assumed that increase in the number of livestock species raised by households' decreases vulnerability. By taking the inverse, we created an index that, for example, assigns greater vulnerability for households with a low number of livestock species. The expression for the livestock diversity index is [1/(number of livestock species + 1)]. The maximum and minimum values were also transformed following this logic, and Eq. (7.1) was then used to standardize this sub-component. After this standardization, an index for each major component of vulnerability was created by the following formula:

$$M_{di} = \frac{\sum_{i=1}^{n} Index_{s_d i}}{n} \tag{7.2}$$

Here M_{di} is the ith component for district d, $Index_{S_{di}}$ stands for ith sub-component for district d, and n is the number of sub-component. After calculating the values for each of the seven major components for a district, they were averaged using Eq. (7.3) to obtain the district-level LVI:

$$LVI_d = \frac{\sum_{i=1}^{7} W_{M_i M_{di}}}{\sum_{i=1}^{7} W_{M_i}} \tag{7.3}$$

where LVI_d, the Livelihood Vulnerability Index for district d, equals the weighted average of the seven major components. The weights of each major component, W_{Mi}, are determined by the number of sub-components that make up each major component and are included to ensure that all sub-components contribute equally to the overall LVI (Sullivan et al., 2002). The LVI is scaled to range from 0 (least vulnerable) to 0.5 (most vulnerable).

7.2 Livelihood Vulnerability Index (LVI) for Tripura—Composite Index Approach

The estimated LVI for the state of Tripura is presented in Table 7.3 showing sub-component standardised values for each district. The major components and the composite LVI for each section are presented in Table 7.4. The dependency ratio index was higher for South Tripura (0.438) and lowest for Unakoti (0.100). Overall, however, Dhalai showed greater vulnerability on the Socio-Demographic Profile index, and Khowai showed the lowest vulnerability on the Socio-Demographic

Profile ($SDP_{Dhalai} = 0.411$, $SDP_{Khowai} = 0.281$). Dhalai respondents reported a higher proportion of female-headed households and a smaller proportion of household heads that attended school. Often respondents listed the older adult in the family as the head of the home, but when questioned further, explained that he/she is not earning any bread for the family presently. In these cases, the member who makes the central part of family income was coded as the head of the household. The average age of household head is observed to be lowest in Khowai and highest in Gomati district. Dhalai also showed greater vulnerability on the livelihood strategies component (0.504) while West Tripura the lowest vulnerability (0.395). More than 90% of the households in all eight districts are reported to be dependent on agriculture and allied activities. However, a higher percentage of households in North Tripura and Sepahijala reported relying solely on agriculture for income. Similarly, agricultural Livelihood Diversification Index is observed to be highest in Sepahijala and Dhalai respectively indicating that these districts are less diversified in terms of livelihood strategies which include growing crops, raising animals, and collecting natural resources.

Time used to travel to a health facility is reported to be highest in Dhalai district (135 min) marginally followed by Gomata (127 min), Khowai (115 min) and Unakoti (112 min). On the other hand, households in North Tripura district (81 min) spend relatively lower time in traveling a health facility. Chronic illness was reported by 17% of households in Sepahijala compared to 5% in the state on average. Similarly, around 22% of households in Dhalai district said that a family member missed work or school due to illness in the past 2 weeks compared to 16% of state average. Unakoti households reported of being more vulnerable to malaria than other districts of the state. When the sub-components were combined, the overall health vulnerability score was observed to be highest for Sepahijala (0.366) and lowest for West Tripura.

There is also variation in social networks indicators among the districts. None of the households in Dhalai and Unakoti districts approached their local government for assistance in the past month. Households of Unakoti district reported borrowing money more frequently relative to the number of times they lent money. People of Sepahijala are observed to receive relatively more in-kind assistance from family and friends Overall, households in Dhalai (0.849) and Unakoti (0.836) were found to be more vulnerable on the Social Networks component.

More than 90% of households in all eight districts depend solely on family farm for food. Around 85% of the households on average reported struggling to find adequate food for their families in the state. This figure is reported highest for Unakoti and lowest for Dhalai. Similarly, households in West Tripura are found to have higher crop diversification compared to state average. Finally, percentage of households that do not save crops and seeds are found to highest in Dhalai district. The overall Food Vulnerability is found to be highest in Khowai and lowest in West Tripura.

All the respondents in Dhalai, Khowai and Unakoti have reported to use a natural water source. Around 11% of the households in the states have reported water conflict. However, this figure was significantly lower for North Tripura (6%). Our analysis shows that considerable time (89 min) is spent in collecting water for household use. However, the figure is much lower for districts like West and South Tripura. Overall,

Table 7.3 Estimated value of sub components of LVI across districts in Tripura

Major components	Sub-components	Districts							
		Dhalai	Gomati	Khowai	North Tripura	Sepahijala	South Tripura	Unakoti	West Tripura
1	2	3	4	5	6	7	8	9	10
Socio-demographic profile	DR	0.425	0.250	0.114	0.200	0.271	0.438	0.100	0.250
	FH	0.150	0.075	0.029	0.050	0.025	0.013	0.075	0.079
	Age	0.450	0.590	0.420	0.480	0.580	0.556	0.456	0.512
	SCH	0.620	0.520	0.560	0.570	0.530	0.550	0.550	0.420
Livelihood	WDC	0.210	0.180	0.150	0.210	0.180	0.190	0.110	0.140
	SA	0.951	0.924	0.971	0.975	0.976	0.937	0.981	0.916
	LDI	0.234	0.125	0.102	0.172	0.238	0.196	0.218	0.102
Social networks	RGR	0.784	0.741	0.728	0.561	0.821	0.594	0.691	0.483
	BLR	0.762	0.739	0.491	0.528	0.682	0.576	0.817	0.452
	NGA	0.100	0.923	0.820	0.775	0.891	0.870	0.100	0.718
Health	ATH	0.781	0.681	0.814	0.478	0.735	0.371	0.691	0.281
	MCI	0.075	0.025	0.070	0.020	0.175	0.019	0.062	0.012
	MWS	0.221	0.150	0.125	0.117	0.181	0.166	0.200	0.160
	MEI	0.210	0.280	0.250	0.110	0.371	0.151	0.330	0.116
Food	AD	0.970	0.960	0.970	0.980	0.950	0.960	0.970	0.950
	SF	0.050	0.070	0.120	0.060	0.090	0.080	0.130	0.070
	CDI	0.110	0.210	0.170	0.280	0.220	0.160	0.260	0.190
	NSC	0.150	0.150	0.170	0.060	0.130	0.060	0.120	0.090
	NSS	0.130	0.120	0.150	0.040	0.120	0.030	0.070	0.040
Water	WC	0.080	0.130	0.170	0.060	0.110	0.080	0.160	0.080

(continued)

Table 7.3 (continued)

Major components	Sub-components	Districts							
		Dhalai	Gomati	Khowai	North Tripura	Sepahijala	South Tripura	Unakoti	West Tripura
1	2	3	4	5	6	7	8	9	10
	NWS	1.000	0.750	1.000	0.975	0.825	0.639	1.000	0.700
	TWS	0.781	0.656	0.579	0.351	0.518	0.295	0.621	0.221
	NCW	0.050	0.070	0.110	0.080	0.130	0.070	0.130	0.100
Natural disasters and climate variability	ND	0.000	0.000	0.000	0.231	0.314	0.371	0.000	0.161
	ID	0.000	0.000	0.000	0.010	0.020	0.000	0.000	0.000
	SDXT	0.017	0.018	0.017	0.018	0.018	0.018	0.018	0.017
	SDMT	0.020	0.201	0.203	0.022	0.200	0.021	0.202	0.020
	SDPR	0.498	0.531	0.519	0.458	0.471	0.561	0.481	0.466

Source Calculated by the author

Table 7.4 Estimated value for major components and LVI across districts in Tripura

Major components	District							
	Dhalai	Gomati	Khowai	North Tripura	Sepahijala	South Tripura	Unakoti	West Tripura
Socio-demographic profile (SDP)	0.411	0.359	0.281	0.325	0.352	0.389	0.295	0.315
Livelihood	0.504	0.437	0.446	0.482	0.481	0.468	0.465	0.395
Social network	0.849	0.801	0.680	0.621	0.798	0.680	0.836	0.551
Health	0.322	0.284	0.315	0.181	0.366	0.177	0.321	0.169
Food	0.298	0.302	0.325	0.255	0.302	0.258	0.310	0.253
Water	0.478	0.402	0.465	0.367	0.396	0.271	0.478	0.275
Natural disasters and climate variability	0.107	0.150	0.148	0.148	0.205	0.194	0.140	0.133
Livelihood vulnerability index (LVI)	0.393	0.365	0.358	0.320	0.387	0.329	0.378	0.280

Source Calculated by the author

Dhalai and Unakoti are observed to have relative higher water vulnerability that n other districts.

In respect of natural disasters, households in Dhalai, Gomati and Khowai districts reported that they did not experience any flood, drought, and cyclone events in the past 10 years and, hence, no injury or death is reported in these three districts because of natural disasters. When climate variability was integrated into Natural Disaster Index however, households in Sepahijala were found to be more vulnerable than other districts.

Overall, Dhalai (0.393) and Sepahijala (0.387) had a higher LVI indicating relatively greater vulnerability of mixed agro-livestock small-holders of these districts to climate change. West Tripura is observed to be the least vulnerable district in the state as per our analysis. The results for livelihood vulnerably among the districts are presented in a spider diagram (Chart 7.1).

Chart 7.1 Variation in LVI across districts in Tripura

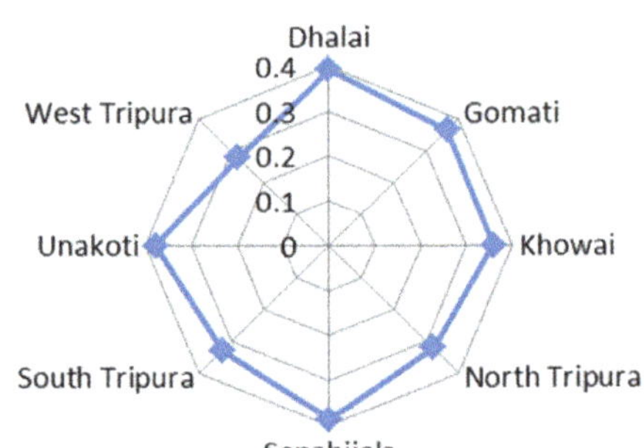

7.3 Livelihood Vulnerability Index (LVI) in Mizoram—Composite Index Approach

The data in Table 7.5 presents the LVI sub-component standardized values for each district of Mizoram. The major components and the composite LVI for each district are presented in Table 7.6. The Dependency Ratio Index was higher for the district Serchhip (0.232) and lowest for Aizawl (0.159). Overall however, Lunglei showed greater vulnerability on the Socio-Demographic Profile index and Aizawl showed the lowest vulnerability on the Socio-Demographic Profile ($SDP_{Lunglei} = 0.395$, $SDP_{Aizawl} = 0.259$). Lunglei respondents reported a higher proportion of female-headed households and Serchhip respondents reported a smaller proportion of household heads that attended school. Often respondents listed the most elderly person in the family as the head of the household but when questioned further, explained that he/she is not earning any bread for the family presently. In these cases, the member who earns the major part of family income was coded as the head of the household. The average age of household head is observed to be lowest in Aizawl and highest in Mamit.

On the Livelihood Strategies component, the district of Mamit (0.530) showed greater vulnerability, while Aizawl showed the lowest vulnerability (0.487). More than 90% of the households in all eight districts are reported to be dependent on agriculture and allied activities. However, a higher percentage of households in Kolasib and Mamit reported relying solely on agriculture for income. Similarly, agricultural Livelihood Diversification Index is observed to be highest in Mamit and Serchhip respectively indicating that these districts are less diversified in terms of livelihood strategies which include growing crops, raising animals, and collecting natural resources.

Time used to travel to a health facility is reported to be highest in Serchhip (155 min) marginally followed by Mamit (149 min). On the other hand, households in Aizawl (45 min) spend relatively lower time in travelling a health facility. Chronic illness was reported by 25% of households in Serchhip compared to 11% in the state on average. Similarly, around 24% of households in Mamit district said that a family member missed work or school due to illness in the past two weeks compared to 17% of the state average. The households in the district of Mamit also reported being more vulnerable to malaria than other districts of the state. When the sub-components were combined, the overall Health vulnerability score was observed to be highest for Serchhip (0.397) and lowest for Aizawl (0.179).

There is also variation in social networks indicators among the districts. More than 90% of the households in Mamit and Serchhip did not approach their local government for assistance in the past month. Households of Serchhip district reported borrowing money more frequently relative to the number of times they lent money. People of Aizawl are observed to receive relatively more in-kind assistance from family and friends. Overall, households in Serchhip (1.136) and Mamit (1.091) were found to be more vulnerable on the Social Networks component.

Table 7.5 Estimated value of sub components of LVI across districts in Mizoram

Major components	Sub-components	Districts						
		Aizawl	Champhai	Kolasib	Lawngtlai	Lunglei	Mamit	Serchhip
Socio-demographic profile	DR	0.159	0.172	0.176	0.190	0.197	0.218	0.232
	FH	0.216	0.350	0.529	0.648	0.727	0.517	0.136
	Age	0.481	0.484	0.492	0.510	0.531	0.544	0.540
	SCH	0.178	0.139	0.104	0.140	0.125	0.131	0.195
Livelihood	WDC	0.120	0.160	0.190	0.19	0.2	0.22	0.22
	SA	0.956	0.961	0.981	0.951	0.924	0.971	0.975
	LDI	0.924	0.951	0.956	0.961	0.971	0.975	0.981
Social networks	RGR	0.845	0.945	1.054	1.160	1.050	1.230	1.250
	BLR	0.926	0.956	1.040	1.060	1.030	1.120	1.170
	NGA	0.672	0.775	0.820	0.870	0.891	0.923	0.987
Health	ATH	0.371	0.478	0.681	0.691	0.735	0.781	0.814
	MCI	0.016	0.018	0.075	0.100	0.175	0.190	0.250
	MWS	0.218	0.125	0.150	0.143	0.181	0.241	0.195
	MEI	0.110	0.151	0.250	0.210	0.280	0.371	0.330
Food	AD	0.961	0.952	0.975	0.978	0.981	0.964	0.978
	SF	0.050	0.060	0.070	0.069	0.090	0.134	0.130
	CDI	0.110	0.160	0.170	0.210	0.220	0.260	0.280
	NSC	0.050	0.060	0.120	0.060	0.170	0.150	0.130
	NSS	0.020	0.030	0.040	0.070	0.120	0.120	0.110
Water	WC	0.080	0.050	0.170	0.080	0.110	0.130	0.160
	NWS	0.639	0.871	0.825	0.983	1.000	0.975	1.000

(continued)

Table 7.5 (continued)

Major components	Sub-components	Districts						
		Aizawl	Champhai	Kolasib	Lawngtlai	Lunglei	Mamit	Serchhip
	TWS	0.285	0.351	0.471	0.579	0.621	0.656	0.781
	NCW	0.050	0.070	0.070	0.080	0.110	0.130	0.130
Natural disasters and climate variability	ND	0.000	0.014	0.000	0.231	0.314	0.371	0.015
	ID	0.000	0.000	0.000	0.062	0.004	0.000	0.008
	SDXT	0.019	0.017	0.018	0.018	0.018	0.018	0.018
	SDMT	0.241	0.201	0.203	0.195	0.139	0.151	0.202
	SDPR	0.479	0.561	0.571	0.491	0.471	0.491	0.481

Source Calculated by the author

Table 7.6 Estimated value for major components and LVI across districts of Mizoram

Major components	DISTRICT						
	Aizawl	Champhai	Kolasib	Lawngtlai	Lunglei	Mamit	Serchhip
Socio-demographic profile	0.259	0.286	0.325	0.372	0.395	0.352	0.276
Livelihood	0.487	0.499	0.511	0.523	0.523	0.530	0.529
Social network	0.814	0.892	0.971	1.030	0.990	1.091	1.136
Health	0.179	0.193	0.289	0.286	0.343	0.396	0.397
Food	0.238	0.252	0.275	0.277	0.316	0.326	0.326
Water	0.263	0.338	0.384	0.431	0.460	0.473	0.518
Natural disasters and climate variability	0.148	0.159	0.158	0.199	0.189	0.206	0.145
Livelihood vulnerability index (LVI)	0.313	0.345	0.382	0.411	0.427	0.445	0.434

Source Calculated by the author

More than 90% of households in all eight districts depend solely on the family farm for food. Around 8% of the households on average reported struggling to find adequate food for their families in the state. This figure is reported highest for Mamit and lowest for Serchhip. Similarly, households in Aizawl are found to have higher crop diversification compared to the state average. Finally, percentage of households that do not save crops and seeds are found to highest in Lunglei district. The overall food vulnerability is found to be highest in Mamit and lowest in Aizawl.

All the respondents in Lunglei and Serchhip have reported using a natural water source. Around 10% of the households in the states have reported water conflict. However, this figure was significantly lower for Champhai (5%). Our analysis shows that considerable time (73 min) is spent on collecting water for household use. However, the figure is much lower for districts like Aizawl and Champhai. Overall, Mamit and Serchhip are observed to have relative higher water vulnerability than other districts.

In respect of natural disasters, households in Aizawl and Kolasib reported that they did not experience any flood, drought, and cyclone events in the past ten years and, hence, no injury or death is written in these three districts because of natural disasters. When climate variability was integrated into Natural Disaster index however, households in Mamit were found to be more vulnerable than other districts.

Overall, the Mamit district (0.445) and Serchhip (0.434) had a higher LVI indicating relatively greater vulnerability of mixed agro-livestock smallholders of these districts to climate change. Aizawl is observed to be the least vulnerable district in the state as per our analysis. The results for livelihood vulnerably among the districts are presented in a spider diagram (Chart 7.2).

Chart 7.2 Variation in LVI across districts in Mizoram

7.4 Livelihood Vulnerability Index (LVI) in Meghalaya—Composite Index Approach

The estimated values in Table 7.7 presents the LVI sub-component standardized values for each district in Meghalaya. The major components and the composite LVI for each district are presented in Table 7.8. The Dependency Ratio Index was higher for West Jaintia Hills (0.510) than Ri-Bhoi (0.470). Overall, however, Ri-Bhoi showed greater vulnerability on the Socio-Demographic Profile index than West Jaintia Hills ($SDP_{\text{Ri-Bhoi}} = 0.463$, $SDP_{\text{West Jaintia Hills}} = 0.433$). Ri-Bhoi respondents reported a higher proportion of female-headed households and a smaller proportion of household heads that attended school. Often respondents listed the most elderly person in the family as the head of the household but when questioned further, explained that he/she is not earning any bread for the family presently. In these cases, the member who earns the major part of family income was coded as the head of the household. The average age of household head is observed to be relatively higher in West Jaintia Hills.

Ri-Bhoi also showed greater vulnerability on the Livelihood Strategies component (0.451) than West Jaintia Hills (0.403). More than 90% of the households in both the districts reported to be dependent on agriculture and allied activities. A higher percentage of Ri-Bhoi (0.963) households reported relying solely on agriculture for income Similarly, agricultural Livelihood Diversification Index is observed to be highest in Ri-Bhoi indicating that this district is less diversified in terms of livelihood strategies which include growing crops, raising animals, and collecting natural resources.

Time used to travel to a health facility is reported to be higher in West Jaintia Hills (96 min) than Ri-Bhoi (83 min). Chronic illness was reported by 9% of households in Ri-Bhoi compared to 12% in West Jaintia Hills. Similarly, around 37% of households in Ri-Bhoi district said that a family member missed work or school due to illness in the past 2 weeks compared to 26% in West Jaintia Hills. Ri-Bhoi households reported of being more vulnerable to malaria than other district of the state. When the sub-components were combined, the overall Health vulnerability score was observed to be higher for Ri-Bhoi (0.366).

There is also variation in social networks indicators among the districts. Over 89% of Ri-Bhoi and 76% of West Jaintia Hills households said they had not approached their local government for assistance in the past month. Households of Ri-Bhoi district reported borrowing money more frequently relative to the number of times they lent money. People of Ri-Bhoi are observed to receive relatively more in-kind

Table 7.7 Estimated value of sub components of LVI across districts in Meghalaya

Major components	Sub-components	Districts	
		Ri-Bhoi	West Jaintia Hills
Socio-demographic profile	DR	0.470	0.510
	FH	0.290	0.210
	Age	0.510	0.560
	SCH	0.580	0.530
Livelihood	WDC	0.130	0.160
	SA	0.963	0.916
	LDI	0.261	0.132
Social networks	RGR	0.381	0.317
	BLR	0.681	0.538
	NGA	0.893	0.762
Health	ATH	0.760	0.790
	MCI	0.091	0.120
	MWS	0.371	0.263
	MEI	0.241	0.189
Food	AD	0.973	0.921
	SF	0.003	0.006
	CDI	0.333	0.200
	NSC	0.160	0.130
	NSS	0.130	0.110
Water	WC	0.070	0.040
	NWS	0.916	0.867
	TWS	0.673	0.592
	NCW	0.078	0.046
Natural disasters and climate variability	ND	0.060	0.080
	ID	0.020	0.030
	SDXT	0.017	0.018
	SDMT	0.023	0.024
	SDPR	0.341	0.327

Source Calculated by the author

assistance from family and friends Overall, households in Ri-Bhoi (0.651) were found to be more vulnerable on the Social Networks component.

More than 90% of households in both the districts depend solely on family farm for food. Less than 1% of the households on average reported struggling to find adequate food for their families in the two districts. This figure, however, is reported to be highest in Ri-Bhoi. Similarly, households in West Jaintia Hills are found to have higher crop diversification compared to Ri-Bhoi. Finally, percentage of households

Table 7.8 Estimated value for major components and LVI across districts in Meghalaya

Major components	Ri-Bhoi	West Jaintia H ills
Socio-demographic profile (SDP)	0.463	0.433
Livelihood	0.451	0.403
Social network	0.651	0.538
Health	0.366	0.341
Food	0.320	0.273
Water	0.434	0.386
Natural disasters and climate variability	0.092	0.096
Livelihood vulnerability index (LVI)	0.372	0.332

Source Calculated by the author

that do not save crops and seeds are found to higher in Ri-Bhoi district. The overall Food vulnerability is found to be higher in Ri-Bhoi.

More than 90% respondents in Ri-Bhoi and 86% in West Jaintia Hills have reported to use a natural water source. Around 11% of the households in the study area have reported water conflict. However, this figure was significantly lower for West Jaintia Hills (4%). Our analysis shows that considerable time (57 min) is spent in collecting water for household use. However, the figure is marginally lower for West Jaintia Hills. Overall, West Jaintia Hills observed to have relative lower water vulnerability than Ri-Bhoi.

In respect of natural disasters, very few households in both the districts have experienced any flood, drought, and cyclone events in the past 10 years and, hence, number of injury caused by natural disaster is also very low. When climate variability was integrated into Natural Disaster index however, households in Ri-Bhoi were found to be slightly more vulnerable than West Jaintia Hills.

Overall, Ri-Bhoi district had a higher LVI than West Jaintia Hills (0.372 vs. 0.332, respectively), indicating relatively greater vulnerability to climate change impacts. The results for livelihood vulnerably among the districts are presented in diagram (refer Chart 7.3).

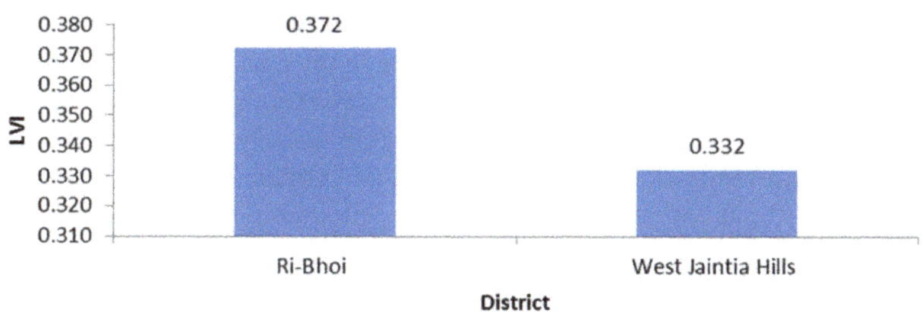

Chart 7.3 Variation in LVI across districts in Meghalaya

7.5 Analysis of the Livelihood Vulnerability Index (LVI) in Arunachal Pradesh—Composite Index Approach

Table 7.9 presents the LVI sub-component standardized values for each district for the state of Arunachal Pradesh. The major components and the composite LVI for each district are presented in Table 7.10. The dependency ratio index was higher for Kurum Kumey (0.481) than Dibang Valley (0.420). Overall, however, Kurum Kumey showed greater vulnerability on the Socio-Demographic Profile index than Dibang Valley ($SDP_{\text{Kurum Kumey}} = 0.597$, $SDP_{\text{Dibang Valley}} = 0.437$). Kurum Kumey respondents reported a higher proportion of female-headed households and a smaller proportion of household heads that attended school. Often respondents listed the most elderly person in the family as the head of the household but when questioned further, explained that he/she is not earning any bread for the family presently. In these cases, the member who earns the major part of family income was coded as the head of the household. The average age of household head is observed to be relatively higher in Kurum Kumey.

Kurum Kumey also showed greater vulnerability on the Livelihood Strategies component (0.479) than Dibang Valley (0.400). More than 90% of the households in both the districts reported to be dependent on agriculture and allied activities. A higher percentage of Kurum Kumey (0.972) households reported relying solely on agriculture for income similarly, agricultural Livelihood Diversification Index is observed to be highest in Kurum Kumey indicating that this district is less diversified in terms of livelihood strategies which include growing crops, raising animals, and collecting natural resources.

Time used to travel to a health facility is reported to be relatively lower in Dibang Valley (78 min) than Kurum Kumey (96 min). Chronic illness was reported by 8% of households in Kurum Kumey compared to 11% in Dibang Valley. Similarly, around 27% of households in Kurum Kumey district said that a family member missed work or school due to illness in the past 2 weeks compared to 19% in Dibang Valley. Kurum Kumey households reported of being more vulnerable to malaria than other district of the state. When the sub-components were combined, the overall Health vulnerability score was observed to be higher for Kurum Kumey (0.341).

There is also variation in social networks indicators among the districts. Over 87% of Kurum Kumey and 78% of Dibang Valley households said they had not approached their local government for assistance in the past month. The households of Kurum Kumey district reported borrowing money more frequently relative to the number of times they lent money. People of Kurum Kumey are observed to receive relatively more in-kind assistance from family and friends Overall, households in Kurum Kumey (0.628) were found to be more vulnerable on the Social Networks component.

More than 90% of households in both the districts depend solely on family farm for food. Less than 1% of the households on average reported struggling to find adequate food for their families in the two districts. Similarly, households in Dibang Valley are found to have higher crop diversification compared to Kurum Kumey.

Table 7.9 Estimated value of sub components of LVI across districts in Arunachal Pradesh

Major components	Sub-components	Districts	
		Kurum Kumey	Dibang Valley
Socio-demographic profile	DR	0.481	0.420
	FH	0.772	0.324
	Age	0.573	0.531
	SCH	0.563	0.472
Livelihood	WDC	0.192	0.142
	SA	0.972	0.941
	LDI	0.274	0.117
Social networks	RGR	0.362	0.271
	BLR	0.648	0.452
	NGA	0.875	0.781
Health	ATH	0.781	0.743
	MCI	0.081	0.116
	MWS	0.271	0.187
	MEI	0.232	0.174
Food	AD	0.968	0.941
	SF	0.004	0.005
	CDI	0.341	0.261
	NSC	0.173	0.126
	NSS	0.127	0.091
Water	WC	0.060	0.020
	NWS	0.815	0.762
	TWS	0.683	0.531
	NCW	0.068	0.036
Natural disasters and climate variability	ND	0.045	0.038
	ID	0.000	0.000
	SDXT	0.014	0.016
	SDMT	0.024	0.022
	SDPR	0.297	0.295

Source Calculated by the author

Finally, percentage of households that do not save crops and seeds are found to higher in Kurum Kumey district. The overall food vulnerability is found to be higher in Kurum Kumey.

More than 80% respondents in Kurum Kumey and 75% in Dibang Valley have reported to use a natural water source. Around 8% of the households in the study area have reported water conflict. However, this figure was significantly lower for Dibang Valley (2%). Our analysis shows that considerable time (63 min) is spent in collecting

Table 7.10 Estimated value for major components and LVI across districts in Arunachal Pradesh

Major components	Kurum Kumey	Dibang Valley
Socio-demographic profile (SDP)	0.597	0.437
Livelihood	0.479	0.400
Social network	0.628	0.501
Health	0.341	0.305
Food	0.323	0.285
Water	0.407	0.337
Natural disasters and climate variability	0.076	0.074
Livelihood vulnerability index (LVI)	0.383	0.315

Source Calculated by the author

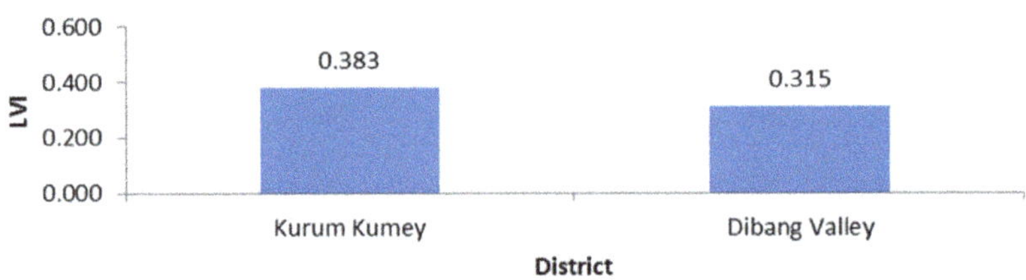

Chart 7.4 Variation in LVI across districts in Arunachal Pradesh

water for household use. However, the figure is marginally lower for Dibang Valley. Overall, Dibang Valley observed to have relative lower water vulnerability than Kurum Kumey.

In respect of natural disasters, very few households in both the districts have experienced any flood, drought, and cyclone events in the past 10 years and, hence, number of injury caused by natural disaster is nil. When climate variability was integrated into Natural Disaster index however, households in Kurum Kumey were found to be slightly more vulnerable than Dibang Valley.

Overall, Kurum Kumey had a higher LVI than Dibang Valley (0.383 vs. 0.315, respectively), indicating relatively greater vulnerability to climate change impacts. The results for livelihood vulnerably among the districts are presented in diagram (Chart 7.4).

7.6 Analysis of the Livelihood Vulnerability Index (LVI) in Manipur—Composite Index Approach

Table 7.11 presents the LVI sub-component standardized values for each district. The major components and the composite LVI for each district are presented in Table 7.12. The dependency ratio index was higher for Churachandpur (0.320) than

Kangpokpi (0.280). Overall, however, Kangpokpi showed greater vulnerability on the Socio-Demographic Profile index than Churachandpur ($SDP_{Churachandpur} = 0.365$, $SDP_{Kangpokpi} = 0.375$). The respondents of Kangpokpi district reported a higher proportion of female-headed households and a smaller proportion of household heads that attended school. Often respondents listed the most elderly person in the family as the head of the household but when questioned further, explained that he/she is not earning any bread for the family presently. In these cases, the member who earns the major part of family income was coded as the head of the household. The average age of household head is observed to be relatively higher in Kangpokpi.

Churachandpur district showed greater vulnerability on the Livelihood Strategies component (0.443) than Kangpokpi (0.419). More than 90% of the households in both the districts reported to be dependent on agriculture and allied activities. A higher percentage of Churachandpur (0.960) households reported relying solely on agriculture for income. Similarly, agricultural Livelihood Diversification Index is observed to be higher in Churachandpur indicating that this district is less diversified in terms of livelihood strategies which include growing crops, raising animals, and collecting natural resources.

Time used to travel to a health facility is reported to be relatively lower in Churachandpur (68 min) than Kangpokpi (76 min). Chronic illness was reported by 7% of households in Churachandpur compared to 6% in Kangpokpi. Similarly, around 24% of households in Churachandpur district said that a family member missed work or school due to illness in the past 2 weeks compared to 19% in Kangpokpi. In respect of malaria hazard, Kangpokpi households reported of being more vulnerable than the other district. When the sub-components were combined, the overall health vulnerability score was observed to be higher for Kangpokpi (0.334).

There is also variation in social networks indicators among the districts. Around 90% of the households in both the districts said they had not approached their local government for assistance in the past month. Households of Kangpokpi district reported borrowing money more frequently relative to the number of times they lent money. People of Kangpokpi are observed to receive relatively more in-kind assistance from family and friends Overall, households in Kangpokpi (0.568) were found to be more vulnerable on the Social Networks component.

More than 95% of households in both the districts depend solely on family farm for food. Less than 1% of the households on average reported struggling to find adequate food for their families in the two districts. Similarly, households in Churachandpur are found to have higher crop diversification compared to Kangpokpi. Finally, percentage of households that do not save crops and seeds are found to higher in Churachandpur district. The overall Food vulnerability is found to be higher in Kangpokpi.

More than 70% respondents in both the districts have reported to use a natural water source. Around 4% of the households in the study area have reported water conflict. Our analysis shows that considerable time (57 min) is spent in collecting water for household use. However, the figure is marginally lower for Churachandpur. Overall, Churachandpur observed to have relative higher water vulnerability than Kangpokpi.

Table 7.11 Estimated value of sub components of LVI across districts in Manipur

Major components	Sub-components	Districts	
		Kurum Kumey	Dibang Valley
Socio-demographic profile	DR	0.320	0.280
	FH	0.170	0.190
	Age	0.528	0.561
	SCH	0.441	0.469
Livelihood	WDC	0.185	0.171
	SA	0.960	0.950
	LDI	0.184	0.137
Social networks	RGR	0.341	0.363
	BLR	0.396	0.426
	NGA	0.894	0.916
Health	ATH	0.735	0.842
	MCI	0.069	0.058
	MWS	0.241	0.195
	MEI	0.225	0.241
Food	AD	0.970	0.980
	SF	0.002	0.003
	CDI	0.316	0.348
	NSC	0.115	0.137
	NSS	0.094	0.118
Water	WC	0.030	0.040
	NWS	0.731	0.784
	TWS	0.693	0.584
	NCW	0.065	0.056
Natural disasters and climate variability	ND	0.000	0.000
	ID	0.000	0.000
	SDXT	0.017	0.015
	SDMT	0.025	0.024
	SDPR	0.273	0.287

Source Calculated by the author

In respect of natural disasters, no households in both the districts have experienced any flood, drought, and cyclone events in the past 10 years and, hence, number of injury caused by natural disaster is nil. When climate variability was integrated into Natural Disaster index however, households in Kangpokpi were found to be slightly more vulnerable than Churachandpur.

Overall, Kangpokpi had a higher LVI than Churachandpur (0.322 vs. 0.327, respectively), indicating relatively greater vulnerability to climate change impacts.

Table 7.12 Estimated value for major components and LVI across districts in Manipur

Major components	Kurum Kumey	Dibang Valley
Socio-demographic profile (SDP)	0.365	0.375
Livelihood	0.443	0.419
Social network	0.544	0.568
Health	0.318	0.334
Food	0.299	0.317
Water	0.380	0.366
Natural disasters and climate variability	0.063	0.065
Livelihood vulnerability index (LVI)	0.322	0.327

Source Calculated by the author

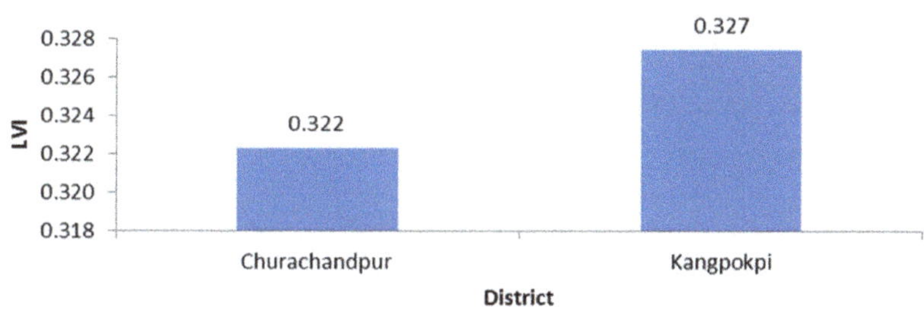

Chart 7.5 Variation in LVI across districts in Manipur

The results for livelihood vulnerably among the districts are presented in diagram (Chart 7.5).

7.7 Analysis of the Livelihood Vulnerability Index (LVI) in Nagaland: Composite Index Approach

The data in Table 7.13 presents the LVI sub-component standardized values for each district in Nagaland. The major components and the composite LVI for each district are presented in Table 7.14. The dependency ratio index was higher for Phek (0.410) than Kohima (0.230). Overall, however, Phek showed greater vulnerability on the Socio-Demographic Profile index than Kohima ($SDP_{Phek} = 0.483$, $SDP_{Kohima} = 0.390$). Phek respondents reported a higher proportion of female-headed households and a smaller proportion of household heads that attended school. Often respondents listed the most elderly person in the family as the head of the household but when questioned further, explained that he/she is not earning any bread for the family presently. In these cases, the member who earns the major part of family income was coded as the head of the household. The average age of household head is observed to be relatively higher in Phek district.

The Phek district also showed greater vulnerability on the Livelihood Strategies component (0.451) than Kohima (0.403). More than 90% of the households in both the districts reported to be dependent on agriculture and allied activities. A higher percentage of Phek (0.963) households reported relying solely on agriculture for income similarly, agricultural Livelihood Diversification Index is observed to

Table 7.13 Estimated value of sub components of LVI across districts in Nagaland

Major components	Sub-components	Districts	
		Kohima	Phek
Socio-demographic profile	DR	0.230	0.410
	FH	0.430	0.480
	Age	0.470	0.530
	SCH	0.430	0.510
Livelihood	WDC	0.110	0.140
	SA	0.960	0.970
	LDI	0.127	0.198
Social networks	RGR	0.294	0.341
	BLR	0.276	0.325
	NGA	0.681	0.812
Health	ATH	0.841	0.873
	MCI	0.084	0.115
	MWS	0.246	0.274
	MEI	0.176	0.215
Food	AD	0.972	0.954
	SF	0.003	0.006
	CDI	0.253	0.326
	NSC	0.127	0.148
	NSS	0.115	0.127
Water	WC	0.050	0.070
	NWS	0.924	0.86
	TWS	0.572	0.625
	NCW	0.054	0.078
Natural disasters and climate variability	ND	0.000	0.000
	ID	0.000	0.000
	SDXT	0.017	0.018
	SDMT	0.025	0.024
	SDPR	0.275	0.315

Source Calculated by the author

Table 7.14 Estimated value for major components and LVI across districts in Nagaland

Major components	Kohima	Phek
Socio-demographic profile (SDP)	0.390	0.483
Livelihood	0.399	0.436
Social network	0.417	0.493
Health	0.337	0.369
Food	0.294	0.312
Water	0.400	0.408
Natural disasters and climate variability	0.063	0.071
Livelihood vulnerability index (LVI)	0.311	0.347

Source Calculated by the author

be highest in Phek indicating that this district is less diversified in terms of livelihood strategies which include growing crops, raising animals, and collecting natural resources.

Time used to travel to a health facility is reported to be lower in Kohima (96 min) than Phek (83 min). Chronic illness was reported by 11% of households in Phek compared to 9% in Kohima. Similarly, around 27% of households in Phek district said that a family member missed work or school due to illness in the past 2 weeks compared to 24% in Kohima. Households of Phek reported of being more vulnerable to malaria than other district of the state. When the sub-components were combined, the overall Health vulnerability score was observed to bc higher for Phek (0.369).

There is also variation in social networks indicators among the districts. Over 68% of Kohima and 81% of Phek households said they had not approached their local government for assistance in the past month.

Households of Phek district reported borrowing money more frequently relative to the number of times they lent money. People of Phek are observed to receive relatively more in-kind assistance from family and friends Overall, households in Kohima (0.417) were found to be less vulnerable on the Social Networks component.

More than 90% of households in both the districts depend solely on family farm for food. Less than 1% of the households on average reported struggling to find adequate food for their families in the two districts. This figure, however, is reported to be slightly higher in Phek. Similarly, households in Kohima are found to have higher crop diversification compared to Phek. Finally, percentage of households that do not save crops and seeds are found to higher in Phek district. The overall Food vulnerability is found to be lower in Kohima.

More than 90% respondents in Kohima and 86% in Phek have reported to use a natural water source. Around 12% of the households in the study area have reported water conflict. However, this figure was marginally lower for Kohima (5%). Our analysis shows that considerable time (48 min) is spent in collecting water for household use. However, the figure is marginally lower for Kohima. Overall, Phek observed to have relative higher water vulnerability than Kohima.

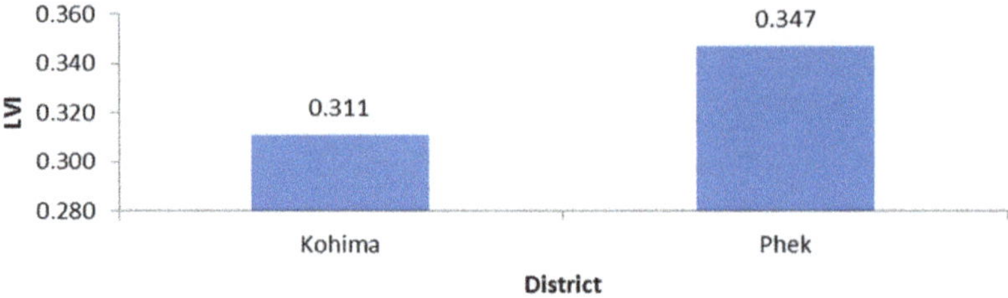

Chart 7.6 Variation in LVI across districts in Nagaland

In respect of natural disasters, none of the households in both the districts have experienced any flood, drought, and cyclone events in the past 10 years and, hence, number of injury caused by natural disaster is nil. When climate variability was integrated into Natural Disaster index however, households in Phek were found to be slightly more vulnerable than Kohima.

Overall, Phek had a higher LVI than Kohima (0.347 vs. 0.311, respectively), indicating relatively greater vulnerability to climate change impacts. The results for livelihood vulnerably among the districts are presented in diagram (Chart 7.6).

7.8 Analysis of the Livelihood Vulnerability Index (LVI) in Assam: Composite Index Approach

Table 7.15 presents the LVI sub-component standardized values for each district. The major components and the composite LVI for each district are presented in Table 7.16. The dependency ratio index was higher for Dima Hasao (0.41) than Karbi Anglong (0.237). Overall, however, Dima Hasao showed greater vulnerability on the Socio-Demographic Profile index than Karbi Anglong ($SDP_{\text{Karbi Anglong}} = 0.402$, $SDP_{\text{Dima Hasao}} = 0.407$). Dima Hasao respondents reported a higher proportion of female-headed households and a smaller proportion of household heads that attended school. Often respondents listed the most elderly person in the family as the head of the household but when questioned further, explained that he/she is not earning any bread for the family presently. In these cases, the member who earns the major part of family income was coded as the head of the household. The average age of household head is observed to be relatively higher in Karbi Anglong.

Dima Hasao showed greater vulnerability on the Livelihood Strategies component (0.475) than Karbi Anglong (0.447). More than 90% of the households in both the districts reported to be dependent on agriculture and allied activities. A higher percentage of Dima Hasao (0.97) households reported relying solely on agriculture for income. Similarly, agricultural Livelihood Diversification Index is observed to be higher in Dima Hasao indicating that this district is less diversified in terms of livelihood strategies which include growing crops, raising animals, and collecting natural resources.

Table 7.15 Estimated value of sub components of LVI across districts in Assam

Major components	Sub-components	Districts	
		Karbi Anglong	Dima Hasao
Socio-demographic profile	DR	0.370	0.410
	FH	0.170	0.140
	Age	0.590	0.560
	SCH	0.478	0.519
Livelihood	WDC	0.120	0.140
	SA	0.950	0.970
	LDI	0.271	0.315
Social networks	RGR	0.275	0.325
	BLR	0.562	0.617
	NGA	0.914	0.948
Health	ATH	0.763	0.816
	MCI	0.075	0.095
	MWS	0.190	0.230
	MEI	0.256	0.185
Food	AD	0.945	0.963
	SF	0.004	0.005
	CDI	0.281	0.361
	NSC	0.195	0.175
	NSS	0.148	0.126
Water	WC	0.060	0.040
	NWS	0.75	0.821
	TWS	0.487	0.534
	NCW	0.067	0.074
Natural disasters and climate variability	ND	0.071	0.089
	ID	0.000	0.000
	SDXT	0.019	0.017
	SDMT	0.024	0.026
	SDPR	0.451	0.463

Source Calculated by the author

Time used to travel to a health facility is reported to be relatively lower in Dima Hasao than Karbi Anglong. Chronic illness was reported by 9% of households in Dima Hasao compared to 7% in Karbi Anglong. Similarly, around 23% of households in Dima Hasao district said that a family member missed work or school due to illness in the past 2 weeks compared to 19% in Karbi Anglong. In respect of malaria hazard, Karbi Anglong households reported of being more vulnerable than the other district.

Table 7.16 Estimated value for major components and LVI across districts in Assam

Major components	Karbi Anglong	Dima H asao
Socio-demographic profile (SDP)	0.402	0.407
Livelihood	0.447	0.475
Social network	0.584	0.630
Health	0.321	0.332
Food	0.315	0.326
Water	0.341	0.367
Natural disasters and climate variability	0.113	0.119
Livelihood vulnerability index (LVI)	0.339	0.357

Source Calculated by the author

When the sub-components were combined, the overall Health vulnerability score was observed to be higher for Dima Hasao (0.332).

There is also variation in social networks indicators among the districts. Around 90% of the households in both the districts said they had not approached their local government for assistance in the past month. Households of Dima Hasao district reported borrowing money more frequently relative to the number of times they lent money. People of Dima Hasao are observed to receive relatively more in-kind assistance from family and friends Overall, households in Karbi Anglong (0.584) were found to be less vulnerable on the Social Networks component.

More than 95% of households in both the districts depend solely on family farm for food. Less than 1% of the households on average reported struggling to find adequate food for their families in the two districts. Similarly, households in Karbi Anglong are found to have higher crop diversification compared to Dima Hasao. Finally, percentage of households that do not save crops and seeds are found to higher in Dima Haso district. The overall Food vulnerability is found to be higher in Dima Hasao.

Less than 10% respondents in both the districts have reported to use a natural water source. Around 5% of the households in the study area have reported water conflict. Our analysis shows that considerable time (69 min) is spent in collecting water for household use. However, the figure is marginally lower for Karbi Anglong. Overall, Dima Hasao observed to have relative higher water vulnerability.

In respect of natural disasters, less than 10% households in both the districts have experienced natural disaster in the past 10 years. However, no of injury was reported by any of the household caused by natural disaster. When climate variability was integrated into Natural Disaster index however, households in Dima Hasao were found to be slightly more vulnerable than Karbi Anglong.

Overall, Dima Hasao had a higher LVI than Karbi Anglong (0.357 vs. 0.339, respectively), indicating relatively greater vulnerability to climate change impacts. The results for livelihood vulnerably among the districts are presented in diagram (Chart 7.7).

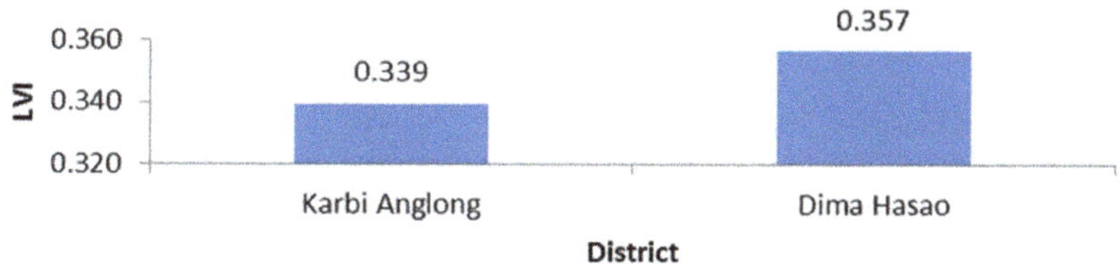

Chart 7.7 Variation in LVI across districts in Assam

In this chapter, we have used composite index approach to estimate the livelihood vulnerability to climate change of people residing in different parts of rural northeastern region of India. Our approach takes into consideration of seven major components: Socio-Demographic Profile, Livelihood Strategies, Social Networks, Health, Food, Water, and Natural Disasters and Climate Variability. Each of these major components comprises several indicators or sub-components.

It has been observed that the values for each component and sub-component of livelihood vulnerability index varied noticeably across districts of NER which provided insight into the design and implementation of district-specific coping strategies for household. As per our analysis, Dhalai in Tripura, Phek in Nagaland, Mamit in Mizoram, Ri-Bhoi in Meghalaya, Kangpokpi in Manipur, Dima Hasao in Assam, and Kurum Kumey in Arunachal Pradesh are the most vulnerable districts in respective state. This observation provides policymakers, planners and development agencies with a scientific instrument to comprehend the role of physical, populations and social factors to climate change vulnerability in different parts of northeast India.

References

Hahn, M. B., Riederer, A. M., & Foster, S. O. (2009). The livelihood vulnerability index: A pragmatic approach to assessing risks from climate variability and change—A case study in Mozambique. *Global Environmental Change, 19*(1), 74–88.

Mainlay, J., & Tan, S. F. (2012). Mainstreaming gender and climate change in Nepal. IIED Climate Change Working Paper No. 2, November, https://www.jstor.org/stable/resrep01234.1.

Panthi, J., Aryal, S., Dahal, P., Bhandari, P., Krakauer, N. Y., & Pandey, V. P. (2016). Livelihood vulnerability approach to assessing climate change impacts on mixed agro-livestock smallholders around the Gandaki River Basin in Nepal. *Regional Environmental Change, 16*(4), 1121–1132.

Shah, K. U., Dulal, H. B., Johnson, C., & Baptiste, A. (2013). Understanding livelihood vulnerability to climate change: Applying the livelihood vulnerability index in Trinidad and Tobago. *Geoforum, 47*, 125–137.

Sullivan, C., Meigh, J. R., & Fediw, T. S. (2002). *Derivation and testing of the water poverty index phase 1*. Final Report. Department for International Development, UK.

Chapter 8
Lessons from the Khonoma Village in India

8.1 Locating Khonoma Village

It is evident from the analysis in the previous chapters that the shifting cultivation, which is locally known as Jhum cultivation, is mostly confined to the tropical rainforests characterised with substantial precipitation, subsistence economy and warm weather. The tribal inhabitants of Northeastern India have long used Jhum, or shifting farming, as a method of subsistence on the hills. The Government of India has recently undertaken a number of initiatives intended to discourage shifting agriculture in the region. One such initiative is the promotion of terrace cultivation with community involvement among the indigenous societies in the region. The indigenous tribal farmers are aware of the environmental consequences of the age old practice of shifting cultivation and its effect on climate change. In the field survey, it was revealed that many of the farmers take initiative in adaptation and mitigation strategies of climate change through discouraging primitive shifting cultivation practices with a sustainable approach by settled methods of cultivation through terracing.

The village Khonoma is one of the sample village selected for conducting field survey in the state of Nagaland, which is situated at the southern part of the state of Nagaland 20 km from the capital city Kohima. The Angami-Naga tribe is the dominant group residing in the village. The village inhabits a land area of around 123 km^2. Because of the efforts of the local Angami Naga-tribe to conserve the environment, Khonoma village has gained recognition as the first "Green Village" in India. The majority of Khonoma's population depends on the land and forest for their daily necessities, agriculture being the prime occupation. With this understanding, they have adapted practices and customs which have helped them to utilise their land and forest sustainably. There are three dwellings in the village called Khels by the locals: Thevomia, Merhüma, and Semoma Khel. There are 424 households and 1943 people living in the village according to the Census Report (2011), Govt. of India. The Gender distribution of the working population in the Khonoma village as per the

N. Roy et al., *Climate Change, Shifting Cultivation and Livelihood Vulnerabilities in India*, https://doi.org/10.1007/978-3-031-54927-4_8

Census (2011) data shows almost equal number of both male and female workforce distribution (refer Chart 8.1).

The literacy rate shows that 53% male is literate as compared to 47% of female population. The distribution workforce between Main and Marginal works shows that 28% are main workers and the majority of the population depend on marginal works. This shows uncertainties in income-earning opportunities of the population in the village (refer Chart 8.2).

Among the Main workers 54.5% are cultivators, 42.7% followed by a few dependent on household industries (refer Chart 8.3).

It is to be noted that the Census of population was conducted by the Govt. of India in (2011) and the field visit in the village occurs in 2018. To understand the latest situation, a primary survey was conducted on 60 households randomly selected from the village during August 2018. The main purpose of the sample survey is to know the pattern of change that has occurred in the methods of shifting cultivation practices. A

Chart 8.1 Gender distribution of working population in Khonoma village. *Source* Census Report (2011), Govt. of India

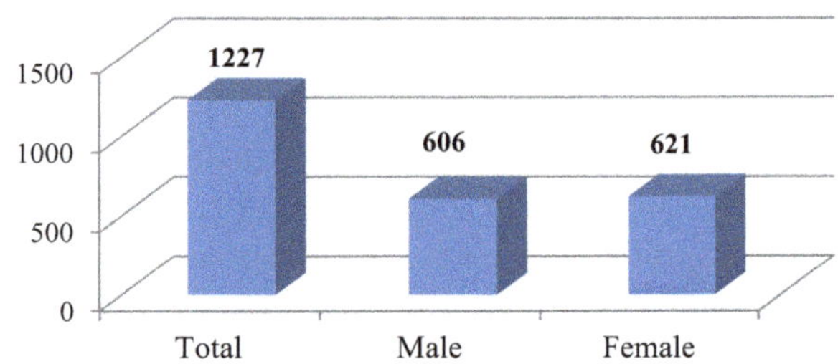

Chart 8.2 Percentage distribution of main and marginal worker in Khonoma village. *Source* Census Report (2011), Govt. of India

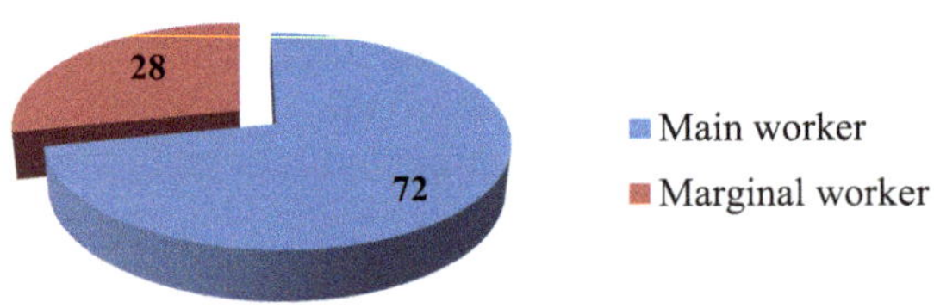

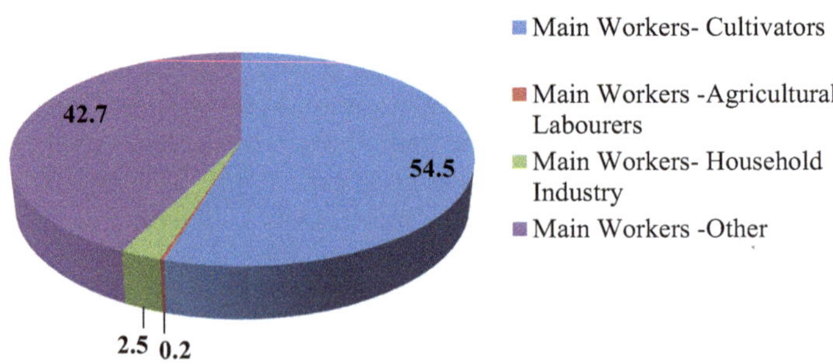

Chart 8.3 Percentage distribution of different categories of main worker. *Source* Census Report (2011), Govt. of India

structured questionnaire was used for the household surveys and includes questions on the underlying drivers of shifting cultivation, including the pattern of livelihood dependency. Further Focused Group discussions and interviews were conducted with various stakeholders like village headman, members of Village Council, members of multiple organisations instituted under the aegis of the Village Council and the public. Data were analysed using simple statistical and graphical tools.

Authors in Discussion with Village Head Men

8.2 Livelihood Pattern in Khonoma

Based on information from the Census report of (2011) as reported in Sect. 8.1 it is revealed that majority of the working population are agriculturists. It has been confirmed by the findings of the field survey. It appears that the majority of Khonoma villagers make a living directly or indirectly from agriculture. It consists of rice cultivation in the surrounding moist terraced rice fields and agricultural products such as vegetables, fruits and cash crops in the jhum fields. In addition to farming, Khonoma people supplement their income by engaging in secondary occupations such as basket weaving, carpentry, wood-working, and stone carving. Some members of the community supplement their income by working in various capacities in government and private institutions. Chart 8.4 shows the background of the main occupations practiced by the sample respondents in Khonoma village.

As seen in Chart 8.4, a total of 65% of the respondents claimed to be employed in agriculture. Thus, agriculture has become the primary means of subsistence for the majority of people in the village, including families and households. There was apparently no major change in the livelihood dependency, as evident from the Census report of (2011). However, field investigations revealed that people mainly produce tiny, scattered fields, partly on lowland valley floors beneath forest peaks, and mostly on

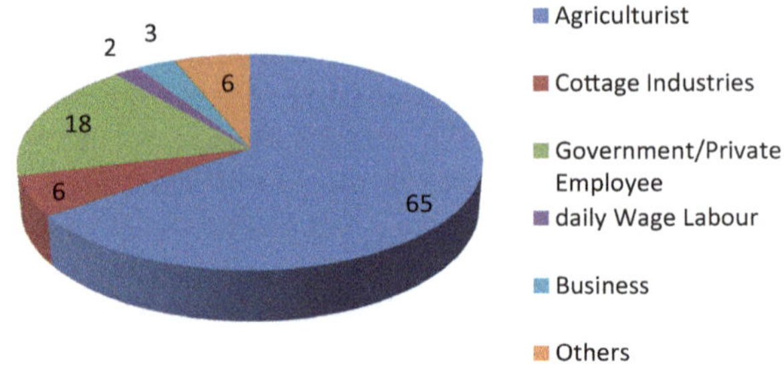

Chart 8.4 Percentage distribution of primary occupation in Khonoma village. *Source* Field survey, 2018

mountain slopes using complementary indigenous irrigation practices and systems. It is demonstrated that they practise rain-fed farming using Aldar tree as nitrogen-fixing plants in the paddy fields. The second largest group, which makes up 18% of the respondents, is made up of government workers such as teachers, school personnel, field workers in the Public Health Engineering and power sectors, and personnel in primary health centres. Most of these people who make a living from cottage industries also sell their goods at retail in the neighbourhood, according to the Focus Group discussions. These individuals can produce large amounts of cottage industry goods, but their output is limited by a lack of funding and market demand. Only 3% of households are self-employed, with the remaining 2% working as day labourers in other people's fields or operating private taxis, general stores, rice mills, sawmills, construction workers etc. Social workers, carpenters, church workers, and retired civil servants were among the respondents who collectively make up 6% of the sample.

It is, however, important to note that the nature of economy is changing in the Khonoma village very fast. From the discussion with the Focused Group, it is observed that though the village has traditionally based on agriculture which has sustained generations, centring on its terrace rice fields and producers from the Jhum fields and forests, the focus of activity in the present days is shifting towards cash crops with emphasis on settled cultivation. Another notable change is that the village has become a tourist destination with the declaration of Khonoma as the first 'Green Village' of India, and the villagers have an alternative means of earning by running home stays and working as tourist guides (KTDB, 2004).

8.3 Cultivation Practices in Knonoma

From the field survey, it is revealed that Khonoma villagers mostly use two types of farming methods: Wet rice cultivation and Jhum cultivation practices. Jhum cultivation is deep-rooted in the traditions of Khonoma people, thereby influencing their ethnic character and community traditions. Alder trees are planted in the fields by the farmers to maintain a sustainable type of Jhum agriculture. In this approach, the land is cultivated for two years, after which it is left fallow for eight to ten years, allowing the Alder trees in the field to naturally replenish the soil's fertility. Alder tree roots stabilise the soil, enhance ground drainage, and so lessen soil erosion. Additionally, Alder tree grows quickly and give the people wood and food, especially in the fallow years. Speaking with the locals also revealed that, especially in the dry seasons, a sizable portion of the households' firewood needs are satisfied by the alder trees planted in the Jhum field. The villagers have been gainfully making use of Alder tree for hundreds of years by mastering the scientific technique of growing and pollarding the alder tree. Alder trees are once more flushed down after harvesting, leaving four to six shoots evenly spaced around the top of the stump head. The technique is then repeated after allowing the shoots to grow until the following Jhum cycle.

Aldar Tree in Khonoma Village, Nagaland

The wet rice cultivation, or "paniketis" as it is known among the local tribes, has two types: wet rice cultivation and wet terraced rice cultivation. Wet rice is grown in lowlands with high rainfall. This is a farming method in which the plot is divided into several sections by creating ridges. This is done to keep the plant partially submerged during part of the year. Rice is grown, followed by wheat, mustard and cold fruits.

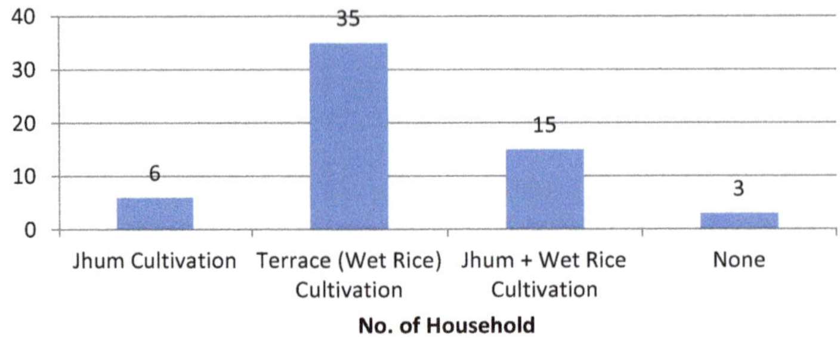

Chart 8.5 Distribution of households in different cultivation methods. *Source* Compiled from field survey, 2018

Plants are sown in sections and watered well. The second method is wet terrace cultivation, in which plants are planted in terraces along the slope. The main crop is rice, but wheat, potatoes, garlic, and cabbage are also grown. Terraced rice farming is permanent and ownership is strictly private. Among the local Naga tribes, the Angami and Chakesang are known for their rice terrace farming.

The data in Chart 8.5 shows that out of the 60 respondents, only 10% of households are currently engaged in Jhum cultivation and rest are engaged in sustainable practices like terrace or combination of both terrace and Jhum cultivation. As revealed by the villagers in group discussions, before the 1970s, most households in the village depended on Jhum cultivation. When villagers were asked why they switched to terraced farming, most said that rice terrace farming is only done during certain months of the year, compared to Jhum farming, which requires working in the fields all year round. Second, unlike Jhum cultivation, paddy cultivation involves permanent human settlement on the cultivated land, so there is no need to clear down forests for cultivation. Additionally, Jhum fields are easy prey for wild animals.

8.4 Production of Crops

Farmers in Konoma traditionally grew red rice, sticky rice, and millet, but production of these crops is now declining. In earlier times, crop production was aimed exclusively at personal consumption and was therefore motivated by household consumption needs. The consumption demand is also dependent on preparing local rice beer and used for religious ceremonies, festivals, and other social ceremonies. However, as social, religious, and economic structures change, the importance of these products is decreasing. The villagers now grow crops not only for their own consumption but also to earn cash income by selling in nearby urban markets.

Transforming Shifting Cultivation to Wet Rice Cultivation in Khonoma, Nagaland

The major crops that villagers grow in the present time in Khonoma include rice, chilli, tomatoes, brinjal, ginger, garlic, pumpkin, cucumber, gourd, yam, lentils, beans, sesame, maize, millet, and job's tears. Thus, cropping decisions are guided by selective cultivation of crops that have higher economic value. This new trend of selective cultivation of cash crops can be said to be because of the change in their economic and social standard. Thus, the farmers are trying to evolve a sustainable cropping pattern which is economically beneficial and also environmentally sustainable. However, this transition of sustainable agricultural practices of shifting cultivators is confined to Khonoma village, as informed by the sample respondents and not practiced nearby villages.

8.5 Perception of the Villagers About Climate Change

From the involvement of the villagers in evolving Aldar tree based cultivation practices over the decades, it is observed that most of them are aware of consequences of the traditional shifting cultivation practices on the environment. Some questions on capturing climate change perceptions among the villagers were asked to understand their knowledge on the environment, climate change and its mitigation practices. When asked whether they had heard the term "climate change," about 93% of participants positively responded that they knew it from the television. However, when asked about their level of awareness of climate change, the majority (65%) of respondents said they were "moderate," followed by 27% "low" awareness and 8% "very aware" of climate change.

The term "climate change" has different meanings for different respondents in the village. About 57% think that it refers to a shift in the pattern or periodicity of precipitation. Some (27%) thought that it would result in fewer droughts or rainfall. However, some (9%) thought that the reason was the rising temperatures. About 4% of the people surveyed indicated decline of soil productivity is linked with climate change. About 80% of respondents stated they had observed changes in the climate in the thirty years prior to the survey. Thus, most of the villagers are aware of climate change and its impact on the agricultural practices.

8.6 Sustaining Shifting Cultivation Practices

Hills' agricultural practices are the main occupation of the villagers in Khonoma. Jhum cultivation is the main source of livelihood for the people living in the village. Despite several institutional efforts, the level of dependency on shifting cultivation has not declined over the years. One of the critical reasons is that the alternatives to Jhum cultivation offered to the cultivators are not easily available to them. The Government of Nagaland enacted the Communitisation of Public Institutions and Service (NCPIS) Act in 2002 to expand the opportunity of community involvement in development and welfare activities in the state. Education, health, electricity, rural water supply, forests, sanitation, and rural childcare were the primary activities that were identified. In order to improve the performance of public utilities, the Government of Nagaland, in collaboration with the tribal communities, undertook management of natural resources through public, private partnership. The emphasis on the development of horticulture has been identified as an important strategy to discourage shifting cultivation practices. Local tribal societies of the Northeastern region in some villages like Khonoma have willingly accepted the sustainable practices after a demonstration of the gains. The Communitisation initiative taken by the Government of Nagaland has a significant impact on awareness and thereby voluntary involvement of the people in the process of socio-economic development.

The Khonoma Village Councils have succeeded in evolving a sustainable method of shifting agriculture practices through a number of initiatives. For example, in Khonoma village as part of Communitisation programme, biodiversity conservation has been undertaken with community initiative. The farmers adopted the sustainable method using an Alder tree in to improve the soil fertility. The majority of the households meet their yearly food from their Jhum farms up to the 1980s. Since the late 1990s, there has been a remarkable transition in Khonoma's agricultural practices towards cash crops due to various programmes undertaken by the state government and implemented by the village council. The Nagaland Empowerment of People through Economic Development programme, earlier known as the Nagaland Environmental Protection and Economic Development (NEPED), also encouraged the farmers to increase their income from 'Jhum fields' by cultivating cash crops such as pepper, turmeric, ginger, and yam in the fallow land (Chawii, 2007).

Summing Up

The experience of Khonoma villagers in sustaining Jhum cultivation and changes in the cropping pattern in favour of cash crops despite difficult topographical structure has been a living example for other shifting cultivators in the Northeastern region. With the cultivation practices becoming settled in nature, the rich biodiversity in Khonoma has led to the declaration of it as "Green Village" of India, which attracted tourists' inflow in the village. Though modest, the villagers have an opportunity for alternative means of livelihood by running home stays and working as tourist guides (KTDB, 2004). Unlike general perception, none of the respondent in Khonoma village considered Jhum cultivation as the cause of climate change. This is because the farmers of Khonoma uphold a sustainable form of Jhum cultivation by planting Alder trees in the fields. In fact, they believe that Alder based shifting cultivation is helpful to maintain a balanced forest ecosystem. Thus, Eco-tourism has opened up a new livelihood opportunity for the villagers.

References

Census. (2011). Govt. of India.

Chawii, L. (2007). Community initiative in biodiversity conservation: Khonoma, Nagaland (Appendix-E). In *Background paper no.14*. Natural Resource-Based Income and Livelihood Improvement Initiatives in Northeast India.

Khonoma Tourism Developement Board (KTDB). (2004). *Environmental Impact Assessment Report*.

Chapter 9
Lessons from the State of Sikkim

9.1 The State of Sikkim—A Profile

Sikkim is the eight state of Northeastern region of India. It borders China in its North and East, Bhutan in its East, Nepal in its West and the Indian state of West Bengal in its South. Sikkim is the least populous and second smallest state in India. The state is a part of the Eastern Himalaya and is prominent for its biodiversity, including Alpine and Sub-tropical climates, it is famous for holding Kanchenjunga, the highest peak in India and third highest on Earth. Almost 35% of the state is covered by the Khangchendzonga National Park. Gangtok is the largest city and capital of Sikkim. The state has unique history of its own before merged to India. Prior to the Indian independence, a resolution in the Indian Constituent Assembly was taken to the effect that Sikkim and Bhutan, were not 'Indian states' and their future should be negotiated separately (Duff, 2015). However, in 1975, the Monarch of Sikkim appealed to the Indian Parliament for Sikkim to become a state of India. In April of that year, the Indian Army took over the city of Gangtok and disarmed the Chogyal's palace guards who ruled Sikkim during 1642–1975. Thereafter, a referendum was held and with the support of 97.5% of voters the system of monarchy was abolished. On 16 May 1975, Sikkim became the 22nd state of the Indian Union. To enable the incorporation of the new state, the Indian Parliament amended the Indian Constitution. First, the 35th Amendment laid down a set of conditions that made Sikkim an "Associate State", a special designation not used by any other state. A month later, the 36th Amendment repealed the 35th Amendment, and made Sikkim a full state, adding its name to the First Schedule of the Constitution of India. The geographical area of the state of Sikkim is 7096 and 2500 km^2 is habitable land.

About a third of the state is heavily forested. The Himalayan Mountains surround the Northern, Eastern and Western borders of Sikkim. The Lower Himalayas, lying in the Southern reaches of the state, are the most densely populated.

The state's economy is largely agrarian based on the terraced farming of rice and the cultivation of crops such as maize, millet, wheat, barley, oranges, tea and

N. Roy et al., *Climate Change, Shifting Cultivation and Livelihood Vulnerabilities in India*, https://doi.org/10.1007/978-3-031-54927-4_9

cardamom. Sikkim produces more cardamom than any other Indian state, and is home to the largest cultivated area of cardamom (India, 2002). Sikkim's Net State Domestic Product (NSDP) was Rs. 3939.82 Crores in 2010–11 at constant (2004–05) prices. Brewing, distilling, tanning and watch making are the main industries, and are mainly located in the southern regions of the state, primarily in the towns of Melli and Jorethang. In addition, a small mining industry exists in Sikkim, extracting minerals such as copper, dolomite, talc, graphite, quartzite, coal, zinc and lead. Despite the state's minimal industrial infrastructure, Sikkim's economy has been among the fastest-growing in India since 2000.

Sikkim is India's least populous state, with 610,577 inhabitants according to the 2011 census with 8.19% below the poverty line. Sikkim is also one of the least densely populated Indian states, with only 86 persons per km^2. However, it has a high population growth rate, averaging 12.4% between 2001 and 2011. The sex ratio is 890 females per 1000 males, recorded in 2011. With around 98,000 inhabitants, the capital Gangtok is the most significant urban area in the state. The total number of household in Sikkim was 129,006 among which 93,288 households were rural inhabitants and only 35,718 were urban as per the Census of 2011. The urbanisation rate in the state was less than 10% and this picture is same for all the four districts in the state. Average household size is 5 and the rural urban decomposition shows a conventional higher value for rural area. The state occupies an area of 7096 km^2 of which 99% is rural area (refer Table 9.1).

Literacy rate in Sikkim was 81.4 which indicate a good picture regarding enrolment in education. The rate varies slightly between rural and urban areas as it is 88.7% for urban and 78.9% for rural area. Literacy rate for male population was 86.6 while it was 75.6 for their other counterpart. It was found to be higher for male for all the districts and also in all the urban areas in all the four districts irrespective of gender (Table 9.2). Another important feature is the concentration of large number Buddhist religious followers in Sikkim. More than half of the population were Hindu (around 58%) according to Census 2011. Second largest population belonged to Buddhism (27%) followed by Christianity (10%). The state also had a little number of Muslim, Sikh and Jain population.

It is found from the Census data that 33.8% of Sikkim's population belonged to Scheduled Tribes (ST) in 2011 while only 4.6% were Scheduled Castes (SC). The pictures varied across districts however, with highest ST population from North Sikkim district (65.7%) and 42% from West Sikkim. South and East Sikkim had almost the same proportion of ST population (Table 9.3).

The analysis of NSDP at factor cost (2004–05 prices) data from Central Statistical Organisation, Govt. of India shows that 57% of state's income accrues to industries while agriculture contributes only 9%. The contribution of service sector is 33% of total NSDP. These figures provides us somewhat different picture from other Northeastern states where industry is not the leading sector of economy.

The data in Table 9.4 deals with distribution of workers in Sikkim. It is found that 95% of urban workers are engaged in other works apart from cultivation, agricultural labour and household industries. However, 48% of rural workers were cultivators. The scenario was quite similar for the districts also. It is also found very limited

Table 9.1 Number of households, total population and household size in Sikkim (Census, 2011)

State/district	Area	Area (in km^2)	No of households	Sex ratio	Population	Household size
1	2	3	4	5	6	7
Sikkim	Total	7096.00	129,006	890	610,577	5
	Rural	7057.75	93,288	882	456,999	4.9
	Urban	38.25	35,718	913	153,578	4.3
North Sikkim	Total	4226.00	8873	767	43,709	4.9
	Rural	4224.24	7819	754	39,065	5
	Urban	1.76	1054	891	4644	4.4
West Sikkim	Total	1166.00	28,023	942	136,435	4.9
	Rural	1162.28	26,890	943	131,187	4.9
	Urban	3.72	1133	937	5248	4.6
South Sikkim	Total	750	30,543	915	146,850	4.8
	Rural	742.4	25,703	908	125,651	4.9
	Urban	7.6	4840	959	21,199	4.4
East Sikkim	Total	954	61,567	873	283,583	4.6
	Rural	928.83	32,876	849	161,096	4.9
	Urban	25.17	28,691	905	122,487	4.3

Source Census (2011), Registrar General of India, Govt. of India

Table 9.2 Literacy rate in Sikkim (Census, 2011)

State/district	Area	Persons	Male	Female
1	2	3	4	5
Sikkim	Total	81.4	86.6	75.6
	Rural	78.9	84.6	72.4
	Urban	88.7	92.4	84.7
North Sikkim	Total	78	83.3	71
	Rural	77.3	82.8	69.9
	Urban	83.8	87.8	79.3
West Sikkim	Total	77.4	83.5	70.9
	Rural	76.9	83.1	70.3
	Urban	89.3	93.5	84.9
South Sikkim	Total	81.4	86.5	75.8
	Rural	80.2	85.6	74.3
	Urban	88.5	92.2	84.6
East Sikkim	Total	83.9	88.5	78.5
	Rural	80	85.5	73.4
	Urban	88.9	92.5	84.9

Source Census (2011), Registrar General of India, Govt. of India

Table 9.3 Social category of population in Sikkim (Census, 2011)

State/district	Total population	SC population	% of SC population	ST population	% of ST population
1	2	3	4	5	6
Sikkim	610,577	28,275	4.6	206,360	33.8
North Sikkim	43,709	982	2.2	28,715	65.7
West Sikkim	136,435	5935	4.4	57,817	42.4
South Sikkim	146,850	6053	4.1	41,392	28.2
East Sikkim	283,583	15,305	5.4	78,436	27.7

Source Census (2011), Registrar General of India, Govt. of India

numbers of engagement into agricultural labourers whereas workers in household industries were very negligible. Thus, this also indicated that Sikkim is not an agrarian economy which further validates not including the state in the sample survey in the present study.

Sikkim has a number of mountain peaks including Kanchenjunga, the world's third highest mountain peak, and attracts a large number of tourists from across the globe. Tourism has developed as the new living of the Sikkimese people with its vast natural prospective. Promotion of village tourism, homestay, cultural tourism,

Table 9.4 Distribution of workers according to economic activities (in percentage)

State/district	Area	Cultivators	Agricultural labourers	Household industries workers	Other workers
1	2	3	4	5	6
Sikkim	Total	38.0	8.0	2.0	52.0
	Rural	48.0	10.0	2.0	40.0
	Urban	1.0	2.0	2.0	95.0
North Sikkim	Total	31.0	10.0	1.0	58.0
	Rural	34.0	10.0	1.0	55.0
	Urban	2.0	4.0	1.0	94.0
West Sikkim	Total	58.0	11.0	2.0	29.0
	Rural	60.0	12.0	2.0	27.0
	Urban	2.0	2.0	3.0	93.0
South Sikkim	Total	51.0	6.0	1.0	42.0
	Rural	56.0	6.0	1.0	36.0
	Urban	3.0	1.0	1.0	95.0
East Sikkim	Total	23.0	8.0	2.0	68.0
	Rural	36.0	12.0	2.0	51.0
	Urban	1.0	2.0	2.0	95.0

Source Census (2011), Registrar General of India, Govt. of India

trekking tourism, ecotourism, wellness tourism, flora-tourism and adventure tourism has given fillip to the tourism trade in the state.

9.2 Cropping Pattern in Sikkim

Rice is one of the main crop in Sikkim as can be inferred from the nickname 'Denzong' meaning 'Valley of Rice'. Maize is ranked second to rice followed by finger millet. Finger Millet or ragi is important minor millet grown in India. In Sikkim finger millet is popularly known as 'Kodo'. The social and economic condition of Sikkim has given a special importance to this crop. Here the grain is mainly used for malting and preparing 'Change' or 'Jansu'. The powder is also used as bread. The stalk is considered to be a good fodder. It is a kharif crop and is cultivated throughout the state irrespective of elevation. The crop occupies about 4.11% of the Gross Cultivated Area. Barley is another important cereal crop of Sikkim. The Green Revolution, which was initiated in the country in the late 1960s, has had a very significant effect in increasing the yield of wheat. The output ratio of wheat to rice has steadily increased 1:3–4:5. Since 1991, the Ministry of Agriculture, Govt. of India has been giving massive thrust to boost its output in the country. Wheat cultivation in non-traditional states including Sikkim is also being popularized by improving irrigation facilities and developing seeds suitable for cultivation in these regions. Pulses locally known as 'Kalodal' or 'Panhelo dal' in Sikkim, is extensively cultivated in all the dry belts of South and West districts. Generally rice and soybean are cultivated during kharif season, so it becomes an optional crop as the farmers has to decide the type of crop to be selected. But in hilly areas of Sikkim where rice production is low, soybean has been profitable crop for these areas. Rape Seed-Mustard is the most important oilseed crop of Sikkim. In Sikkim, this oilseed is cultivated with sole motive of obtaining edible oil as almost all state population utilise this oil as main edible oil, thus it has so much importance.

Sikkim is the large producer of large cardamom and constitute lion share of Indian and world market. Chili is an important cash crop in India and is grown for its pungent fruits, which are used both green and ripe (the latter in the dried form) to impart pungency to the food. Chilly was known to Indians about 400 years ago, when this crop was first introduced into India by Portuguese, towards the end of the fifteenth century. Its cultivation became popular in the seventeenth century. It is now grown in all parts of India covering about 733,800 ha. In Sikkim, cherry pepper also locally known as dale khorsani is the most favorite chili grown in almost every kitchen garden. Ginger locally known as "Adua" is an important spice/cash crop grown in Sikkim since time immemorial. The crop is a good source of income for small and marginal farmers and is grown up to an elevation of 1500 m above msl, occupying an area about 8000 ha producing 44,000 tons of ginger.

Mandarin orange is the most common among citrus fruits grown in India. Sikkim occupies nearly 50% of the total citrus area in India. Sikkim Mandarin represents the most important commercial fruit of Sikkim. Sikkim Mandarin is similar to the

Nepal or Assam or Darjeeling Mandarin. Citrus reticulate is a highly polyembryonic species, having medium sized upright trees. Besides worldwide demand of Mandarin Oranges as fresh fruits, these has been used for the preparation of a number of processed products like orange marmalade, bottled and canned juices, squash, jam, jelly etc. Orange oil is also important citrus oil used for flavoring purposes. Banana is another important fruit and cultivated (hill banana) throughout Sikkim in the kitchen garden, boundaries of field and farms etc. The importance of this as a fruit crop has recently been realised in Sikkim being grown in the foot hills of the state below 900 m elevation. The State Department of Agriculture produces seedlings at Namli garden and Majitar farm in East district and Kaijeley farm in West and distribute to the farmers every year in large scale for the rapid development and growth, of this fruit crop. Although Sikkim receives very high rainfall, which has discouraged the development of this fruit crop, yet there are some pockets in the states such as south district, lower elevation of West districts, and East district with lesser rainfall wherein papaya was successfully cultivated as a commercial fruit crop. It is a highly paying fruit crop with a very high yield potential. Guava is another important fruit crop of the state. Till 1976, it was cultivated in a limited area as in Kitchen garden or as border tree etc. It is now cultivated in the form of orchards below 1200 m elevation in the hills of the states.

Sikkim has the suitable weather for its cultivation of brocolli and, therefore, there is a tremendous scope for its cultivation and supply of broccoli to different parts of the country and even export too. Cauliflower was introduced in India in 1822 (Swarup & Chatterjee, 1972). The initial introduction was the Cornish types Bram England and later the other European types also were introduced. The typical Indian cauliflowers were the result of inter-crossing between the Cornish and other European types. Thus a new development took place in India resulting in tropical cauliflower resistant to hot weather and high rainfall. These types are normally grown in Indian plains from May–June to January which is followed by the only temperate cauliflower known as snowball. All these cauliflowers grown in India are annual, both Indian and snowballs. Among the winter vegetables in India, cauliflower is possibly the most popular. It is an important vegetable crop of Sikkim and is widely cultivated for its white tender head (curd). It is a nutritious vegetable and contains a good amount of Vitamin B and a fair amount of protein, minerals and vitamins A and C.

The area, production and yield of major crops grown in Sikkim are depicted in Table 9.5 for the year 2020–21. The highest area cultivated and production was recorded for maize among the cereals, followed by rice and wheat. Oilseeds have the third rank and that of Pulses fourth rank. Regarding productivity, rice tops the list followed by maize, Barley, wheat and other cereals.

Table 9.5 Area, production and productivity of major crops cultivated in Sikkim during 2020–21

No.	Crop	Area (000' ha)	Production (000' tones)	Productivity (kg/ha)
1	2	3	4	5
1	Rice	8.7	16.19	1860.61
2	Wheat	0.1	0.12	1153.85
3	Maize	38.39	67.94	1769.68
4	Finger-Millet	2.05	2.13	1037.98
5	Barley	0.35	0.4	1167.63
6	Buckwheat	3.11	3.07	986.83
7	Pulses	5.14	4.95	961.76
8	Oilseeds	6.13	5.67	925.09

Source Annual Reports, Agriculture Department, Government of Sikkim Annual Reports, Agriculture Department, Government of Sikkim

9.3 Transition to Organic Farming in Sikkim

In early 2000s, more than 60% state population were dependent on agriculture and allied activities for their livelihood. However, the contribution made by that population to state's economy was only 8%. Thus, understanding the future potential of the agriculture for more inclusive and faster development, Sikkim Legislative Assembly passed a resolution to adopt organic farming in early 2003 (Kumar et al., 2018). Consequently, the Government of Sikkim announced the vision for Sikkim to be India's first organic state in 2003. In 2010, the state launched the 'Organic Mission', an action plan defining the measures to implement in order to reach the target of converting the entire state to organic. Sikkim is the first state in the country to officially adopt and become a totally organic farming state at the end of December 2015. In August 2019, Sikkim was declared as world's first 100% organic state by UN Food and Agriculture (FAO) by converting around 75,000 ha of agricultural land into sustainable cultivation. Agricultural land was gradually converted to certified organic land by implementing organic practices and principles as per guidelines laid down in National Programme for Organic Production. In 2003 the Government of Sikkim decided to make Sikkim an organic farming state through a declaration in the legislative Assembly and the policy implemented a phase-out of chemical fertilisers and pesticides, and achieved a total ban on sale and use of chemical pesticides in the state (Pradhan et al., 2017). The transition has benefitted more than 66,000 farming families. In the long term, organic farming leads in subsistence of agriculture, bio-diversity conservation and environmental protection.

In line with National Programme of Organic Production (NPOP), the 'Organic Mission' prepared a three-phase execution plan to make Sikkim an organic state. Table 9.6 depicts the phase-wise targets and achievements of organic farming area.

Table 9.6 Sikkim organic mission phase-wise targets and achievements

Phase	Financial year	Proposed area (in ha)	Achieved area (in ha)	Achieved (in %)
Phase 1	2010–11	18,000	18,234.33	101.31
Phase 1	2011–12	18,000	19,255.11	106.17
Phase 1	2012–13	14,000	19,283.15	137.73
Total		50,000	56,777.59	113.55

Source Kumar et al. (2018)

As it is found, the total area under organic farming achieved was 56,777.59 ha in the state as against the target of 50,000 ha. Thus the state has always excelled against its targets in all three phases with an average of 113.55% (refer Table 9.6).

During phase 1, 14,232 numbers of farmers were involved in organic farming. In phase 2, the figure has significantly increased to 17,814, and then to 17,333 in phase 3. As organic farming helps to retain and sustain the nutrients of the soil, thus the products produced through such farming retain best of their quality. In addition, financial cost of organic farming is minimum since such farming helps in saving significant amount money which are otherwise spent on artificial fertilisers. Thus, organic farming helps farmers to maintain the complex agro-ecosystem, decreases the risk of health ailments and contributes to ample scope for farming by fetching premium price for the organic products (Kumar et al., 2018).

9.4 Organic Cultivation in Sikkim and Its Impact on Environment and Livelihood

In 2003, Sikkim went on an incredible journey, pledging to become the world's first fully organic state (Paunglad, 2022). This bold commitment was not merely an agricultural shift; rather comprehensive strategy aimed at harmonizing environmental sustainability with the well-being of its farming communities. Over the years, Sikkim's organic cultivation has proved to transform in to a sustainable model, demonstrating the dual impact on environmental conservation and livelihood sustainability.

Environmental Sustainability: Organic farming in Sikkim in the Northeastern region has been a key factor in promoting environmental sustainability through various initiatives:

1. Pesticide-Free Farming

   The main foundational principle of Sikkim's organic farming is the complete elimination of synthetic pesticides and fertilizers (Gregory et al., 2017). This decision was pivotal in the reduction of the environmental impact of agriculture, preventing chemical runoff into water bodies, and protecting the region's delicate ecosystem.

2. Soil Health and Conservation

   Organic farming practices in Sikkim prioritise soil health through the incorporation of organic matter, cover cropping, and reduced tillage. This keeps the soil fertile and inhibits soil erosion, both of which are important in the difficult hilly terrain of the Himalayas. (Chongtham et al., 2019).
3. Biodiversity Preservation

   By avoiding chemical inputs, Sikkim's organic farms have become havens for biodiversity (Babu et al., 2015). Beneficial insects, birds, and microorganisms crucial for natural pest control thrive, creating a balanced and resilient ecosystem.
4. Agroforestry and Carbon Sequestration

   Sikkim has embraced agroforestry, mixing plants with cultivated crops (Subba, 2009). This not only increases biodiversity but also aids in carbon sequestration, mitigating the effects of environmental change and ensuring the long-term viability of the region's land usage. In addition to improving biodiversity, this also helps sequester carbon, reducing the effects of Climate Change and guaranteeing the long-term viability of the area's land use.
5. Water Conservation

   Organic farming practices in Sikkim, such as mulching and cover cropping, enhance water retention in the soil (Mishra & Rai, 2013). This reduces the need for excessive irrigation, aligning with the state's commitment to sustainable water use.
6. Climate Resilience

   Sikkim's commitment to sustainable agriculture incorporates crop diversification, organic farming, and agroforestry, enhancing resilience to climate uncertainties. By embracing these practices, Sikkim not only guarantees food-security but also contributes to global efforts in mitigating climate change impacts on agriculture, creating a model for regions facing similar challenges (Chettri et al., 2019). This all-encompassing strategy fits with Sikkim's goal of having a resilient and sustainable agriculture in the future.

Livelihood Sustainability: The economy of Sikkim underwent a significant transition through the implementation of Organic farming which ensure livelihoods sustainability of the people of the state to a great extent. Some these are:

1. Economic Returns

   The adoption of organic cultivation in the state of Sikkim has resulted in financial gains for farmers. The premium prices that organic produce commands at the marketplace have ensured not just sustenance but improved income for farming communities (Sharma et al., 2020).
2. Market Access

   The commitment to organic cultivation has opened up new markets for Sikkimese farmers. The growing demand for organic goods worldwide has given farmers access to profitable markets, enhancing the agricultural industry's ability to remain financially stable (Tashi et al., 2015).

3. Diversification and Market Stability

   Through the diversification of crops and integrated farming systems, organic farmers in Sikkim have reduced their vulnerability to market fluctuations. This diversification acts as a risk mitigation strategy, ensuring more steady incomes for agriculturalists (Gurung et al., 2018).

4. Community Engagement

   The organic movement in Sikkim has fostered stronger connections between farmers and local communities. Direct sales through farmers' markets and community-supported agriculture initiatives have strengthened community ties, providing mutual support and shared benefits (Tenzing et al., 2019).

5. Health and Safety

   Prioritizing the health and safety of farmers, organic farming practices in Sikkim eliminate the exposure to harmful chemicals associated with conventional agriculture. This ensures the overall well-being of the farming community.

6. Traditional Farming Methods

   The promotion of organic cultivation in Sikkim includes the preservation and promotion of traditional farming methods. This not only ensures the sustainability of cultural practices but also empowers local communities who have deep-rooted knowledge of the land.

7. Knowledge Transfer and Empowerment

   With Sikkim switching to organic farming practise, extensive knowledge transfer initiatives have been put in place. Organic farming training improves farmers' abilities and gives them the knowledge they need to make wise judgements regarding their farming methods.

Sikkim's organic cultivation journey stands as a testament to the transformative power of sustainable agricultural practices. By intertwining environmental stewardship with livelihood sustainability, Sikkim has not only conserved its unique ecosystem but also empowered its farming communities. The achievement of Sikkim's organic mission provides valuable insights and inspiration for regions worldwide seeking to harmonize agricultural productivity with ecological responsibility. The case of Sikkim demonstrates that a holistic and committed approach to organic cultivation can yield positive outcomes, benefitting both the environment and the livelihoods of those dependent on the land.

Summing Up

In this chapter, an attempt has been made to draw lessons from Sikkim about sustainable organic farming. The Government of Sikkim announced the vision for Sikkim to be India's first organic state in 2003. It was the first state in the country to officially adopt and become a totally organic farming state in the end of December 2015.

Organic farming helps farmers to maintain the complex agro-ecosystem, decreases the risk of health ailments and contributes to ample scope for farming by fetching premium price for the organic products. The success story of Sikkim in transformation towards organic agriculture is a lesson for other states in the Northeastern region of India. These states can adopt organic cultivation to a great extent considering the topography and physiographic conditions similar to Sikkim.

References

Babu, S., Singh, R., Avasthe, R. K., & Yadav, G. S. (2015). Organic farming: Problems and prospects in North East India. In *Training manual on integrated farming system approaches for sustainable hill agriculture under changing climatic scenario.* ICAR Research Complex for NEH Region, West Tripura.

Census of India. (2011). Government of India, New Delhi.

Chettri, S., Sharma, E., & Chaudhary, R. P. (2019). Agroforestry in the Eastern Himalayas: A viable and sustainable land use system for mountain development. In K. S. Murali, E. Sharma, & S. Chettri (Eds.), *Agroforestry for biodiversity and ecosystem services: Science and practice* (pp. 77–92). Springer.

Chongtham, I., Sharma, M., & Giri, D. D. (2019). Soil health management through organic farming for sustainable agriculture. In M. Sharma & V. K. Upadhyay (Eds.), *Organic farming for sustainable agriculture* (pp. 255–266). Springer.

Duff, A. (2015). *Requiem for a Himalayan Kingdom.* Birlinn Ltd., Sikkim.

Gregory, L., Plahe, J., & Cockfield, S. (2017). The marginalisation and resurgence of traditional knowledge systems in India: Agro-ecological 'islands of success' or a wave of change? *South Asia: Journal of South Asian Studies, 40*(3), 582–599.

Gurung, A., Tamang, B., Gurung, P., & Tamang, S. (2018). Agriculture and livelihoods in the Sikkim Himalayas: A socioeconomic assessment. *Journal of Agriculture and Environment, 19*, 65–76.

India: A Reference Annual. (2002). Research and Reference Division, Ministry of Information and Broadcasting, New Delhi.

Kumar, J., Pradhan, M., & Singh, N. (2018). Sustainable organic farming in Sikkim: An inclusive perspective. In *Advances in Smart Grid and Renewable Energy* (pp. 367–378). Springer.

Mishra, P. K., & Rai, S. (2013). Use of indigenous soil and water conservation practices among farmers in Sikkim Himalaya. *Indian Journal of Traditional Knowledge, 12*(3), 454–464.

Paunglad, B. (2022). Lesson learned of organic agricultural policy of Sikkim State, India to Thailand. *NIDA Case Research Journal, 14*(2), 27–57.

Pradhan, M., Tripura, B., Mondal, T. K., Darnnel, R., & Murasing, J. (2017). Factors influencing the adoption of organic farming by the farmers of North District of Sikkim. *International Journal of Advanced Scientific Research and Development, 4*(2), 1–7.

Sharma, R., Basumatary, S., & Dutta, M. (2020). Sikkim: A case study on the transition to organic farming. In A. K. Bhattacharya, M. J. Kothari, & C. Pathak (Eds.), *Ecocriticism and environmental communication in South Asia* (pp. 157–171). Springer.

Subba, J. R. (2009). Ingenious agricultural heritage systems and sustainable agricultural ecosystems management in Sikkim. *Bio-Cultural Diveristy & Sustainable Development in North East India: Status, Vision & Challenges*, 86.

Swarup, V., & Chatterjee, S. S. (1972). Origin and genetic improvement of Indian cauliflower. *Economic Botany, 26*, 381–393. https://doi.org/10.1007/BF02860710

Tashi, S., Stellmacher, T., Jamwal, R., & Tsering, D. (2015). Organic farming in Sikkim Himalaya: A successful transition to conservation agriculture. *Mountain Research and Development, 35*(3), 276–287.

Tenzing, T., Wangchuk, D., & Behera, B. (2019). Impact of organic farming on income and employment generation in Sikkim, India. *Indian Journal of Agricultural Economics, 74*(2), 224–235.

Chapter 10
Towards a Sustainable Shifting Agriculture

10.1 Findings from the Analysis of Secondary Data

(1) Apart from Assam and Tripura other Northeastern states are mostly inhabited by indigeneous tribal societies having unique social and cultural practices. About 75% of the Northeastern region is covered by hilly terrain where indigenous communities mainly live. Overall, tribal people account for over 56% of the total population of the region. However, in Meghalaya, Mizoram and Nagaland scheduled tribes comprise more than 80% of the population, the proportion for Arunachal Pradesh is 68%.

(2) Out of eight states in the Northeastern region except Sikkim, shifting cultivation is practiced in all the states in the region. Sikkim has transformed its agricultural sector in to organic farming. It has been observed that out of total liquid based bio-fertiliser in India Sikkim has produced about 90% of production signifying the extent of utilization of organic manures in the state.

(3) Shifting cultivation is the natural way of life of the indigenous societies of the Northeastern region. The system of shifting cultivation is mainly practiced in hilly areas of Northeastern region and in other few states in India. The major advantage of the shifting cultivation to the hill people is that it provides a very easy method for preparing land for cultivation. Weeds and jungles can be easily cleared by slash and burn process and yield can be obtained within a short period of time.

(4) The most important evil effect of shifting cultivation is that destruction of forest in the hill area causes heavy soil erosion due to rainfall. Moreover, loss of the 'top soil' in the hill areas to the extent of 22% of the total soil due to 'jhumming' causes a serious fall in the fertility of the soil and thus it creates a serious problem for the tribal society.

(5) Of the total reporting area, net area sown in the Northeastern region forms only 10.25% whereas at the national level it is about 46.28%. However, the comparison of this nature between the NER and the country as a whole is perhaps not

N. Roy et al., *Climate Change, Shifting Cultivation and Livelihood Vulnerabilities in India*, https://doi.org/10.1007/978-3-031-54927-4_10

proper due to the basic differences in geophysical characteristics. For example 44.20% of the region's reporting area in 2011 was under forests as against 21.29% in the country, which cannot be put to cultivation. The percentage of cultivable area over the geographical area in most of the Northeastern states (26.36%) is less than the national average (32%).

(6) The hills constituted about 70% of the total land area and the low percentage of cultivable land and sparse population in the hill states limits the choices for land based activities in the hills. The difficulties of storing water due to hill slopes have compelled farmers to depend mostly on *jhum* cultivation. The region has a strong advantage in forest based activities and biodiversity, given its high percentage of forest land (72.86%) compared to 21.34% in the country as a whole as per Forest Survey of India report of 2015.

(7) The share of Northeastern region in India's food production has become very insignificant over the years. The technological transformation in agricultural sector started in mid 1960s and resulted in 'Green Revolution' in different parts of the country. This transformation of agricultural sector in India has ensured food security in the entire country. It is only Northeastern region hills could not take part and adapt the technological transformation process till date. The age-old practice of shifting agriculture in traditional tribal societies could not transform in to modern agriculture. Ensuring food security is a big concern for the indigenous household due to geographical remoteness and inaccessibility in many parts in the region.

(8) The agriculture in the Northeastern region is mainly rainfed and only about 10% of total cropped area is under irrigation with wide variation across the states. The rainfall shows wide fluctuations across the eight states in the region. The analysis on the behaviour of annual and seasonal rainfall in Northeastern region of India has been made for understanding trends in long historical data series. A pre-whitening technique has been used to reduce the serial correlation in long time series. In that analysis Sen's coefficient technique has been used to detect the nature of trends of eight rainfall series with more than 100 years of observation. It is observed significant variability in magnitude and direction of monthly rainfall trends from one state to another. Nagaland, Manipur and Assam are the three states for which the trend of rainfall is negative for most of the months. Conversely, Arunachal Pradesh and Meghalaya have experienced increasing trend of rainfall for most of the months. Secondly, seasonal data indicates that, apart from pre-monsoon season, the trend of rainfall is roughly negative in Northeastern India in all other seasons. Year to year variation in rainfall is observed to be the highest in post-monsoon season marginally followed by winter season. Conversely, the year wise variation in rainfall is observed to be the lowest in southwest season. Finally, annual data reveals that Assam, Manipur, Mizoram, Nagaland, Tripura, and Sikkim have experienced decreasing trend of rainfall during the study period. Arunachal Pradesh is the only state where the annual trend of rainfall is on the rise.

(9) The behaviour of annual and seasonal temperature in the Northeastern region of India looking for trends in long historical data series has been analysed. A pre-whitening technique has been used to reduce the serial correlation in long time series. Sen's coefficient technique was used to detect the nature of trends of eight rainfall series with more than 100 years of observation.

It is observed significant rise in temperature across seasons for all the states in NER. However, the rise in temperature is not uniform across the states. Average annual maximum temperature is observed to be highest in Tripura followed by Mizoram and Assam. The figure is lowest for Arunachal Pradesh and Sikkim. On the other hand, average annual minimum temperature is observed to be lowest in Arunachal Pradesh followed by Sikkim and Nagaland. The difference between maximum and minimum temperature is observed to be highest in Meghalaya. In case of seasonal pattern, it is found that Tripura marks the highest maximum temperature in all seasons while Arunachal Pradesh exhibits the lowest temperature in all seasons but southwest monsoon. In southwest monsoon, Nagaland recorded the lowest maximum temperature. The picture is almost similar for minimum average temperature as well. Tripura recorded highest minimum temperature for all seasons baring post monsoon.

(10) The region witnessed loss of forest cover in different assessment of Forest Survey of India (FSI). The per capita availability of forest cover in the region declined sharply over the years. The main reason for this decrease is attributed to increase in population concomitant with the biotic pressure and shifting cultivation in the region. The region witnessed loss of forest cover in different assessment of Forest Survey of India. Consequently per capita availability of forest cover in the NER declined sharply over the years as compared to the country as a whole. While the conservation and social forestry programme of the government both at the centre and states led to increase in forest area in the country as a whole since 1999, no impact has been visible in NER.

(11) The Common Property Resources (CPR) dependent produce play an important role in people's livelihood in the Northeastern region of India. Forest is an important part of CPR. The CPR in the Northeastern region constituted 44.07% of geographical area as against 20.78% in the country as a whole with wide variations amongst the states in 2015–16.

(12) The land holding pattern of the scheduled tribe population of the region has been analysed in the study. It is observed that the distribution of operational holdings in the region is highly skewed with the concentration of 67.9% of the holdings in the marginal and small holding size class. The medium and large holding size in the region constituted 14.5% of the total holdings as against 7.2% for the country as a whole. This may be due to distribution of land by the village headmen to the families on the basis of some ideological norm, say family size. About 80% of the farmers in the NER belongs to the small (< 1.44 ha) and marginal (< 0.40 ha) category. The average size of land holding for the Northeastern states (1.60 ha) is marginally higher than the all India (1.57 ha) average.

(13) There is predominance of cereals, mainly paddy, which accounts for 59.28% of total cropped area in 2015–16 in the Northeastern region. The total food crops constituted 83.65% in the NER as against 71.56% for the country as a whole in 2015–16. This indicates slow process of diversification of agricultural sector in the region. Infact paddy is the main crop in jhumming. It has been observed that about 90% of the total cultivable area in the hills of the Northeastern region belongs to rice.

(14) A high-resolution global potential soil erosion data and global rainfall erosivity data using a conjunction of Remote Sensing and GIS have been analysed in this study. The results clearly indicate the rainfall erosivity was very high in Assam, Meghalaya, Tripura and Mizoram state whereas Arunachal Pradesh, Nagaland and Manipur state has low and moderate rainfall erosivity. The analysis indicated the Assam has associated with high rainfall erosivity as well as high soil erosion. Mizoram has also moderate to high soil erosion whereas rainfall erosivity is also high.

The spatial patterns in land use changes over the time period that directly affects the magnitude of soil erosion which can affect the land productivity and food security, biological diversity and carbon cycling. The Global soil erosion datasets were analysed at 25 km^2 spatial resolution. The MODIS based MCD12Q1 datasets were used to analysed the land use land cover observation in Northeast regions of India. High variation is found in Evergreen Broadleaf forests, savannas, cropland, urban and Built-up lands, water bodies. The observation of the analysis represents the decrease in croplands from 2001 to 2019. The results also represent that the croplands have increased from 2001 to 2010 after that it started declining. Since urbanization and Built-up land are also increasing over the time period, a huge decrease in barren land is observed from 2001 to 2019. It was also observed the tiny fluctuation in soil erosion from 2001 to 2012. Assam state received more rainfall that rain feed the Brahmaputra river. Every year in Assam recurring flood and erosion causes huge loss of life and property, agricultural land being destroyed and unbalance the socio-economic conditions of the state. As well as the results clearly represents the high rainfall erosivity in Assam whereas soil erosion impact is also high.

10.2 Findings from the Analysis of Sample Data

(1) The analysis of the occupational distribution of sample data revealed that about 79% of households of the sample data are engaged in shifting cultivation in the Northeastern region. The transformation in rural livelihood undergoes encouraging trend for the state of Meghalaya. In this state, about 24% people in the sample households are dependent on self-employment in non-farm activities. It is also to be noted that 54.7% of sample households are living below the poverty line.

(2) From the analysis of the distribution of income earned by sample households from different sources, it is observed that about 79% of sample households are engaged in agriculture earning about 47% of income. The analysis also revealed that 24% of income is earned by only 8.6% self-employed in non-farm activities. The highest percentage of income, i.e. 76%, was made from agriculture in the state of Nagaland and the lowest in Arunachal Pradesh.
(3) From the analysis it has been observed that for rice, about 89% area covered by the sample households in the state of Mizoram and lowest, i.e. 39.5% in the state of Meghalaya. The household-level landholding pattern for rice cultivation shows 2.3 acres in the region as a whole. In comparison, the sample farmers in the state of Arunachal Pradesh allocate 3.0 acres for rice and that of lowest i.e. 1.4 acres in Meghalaya.
(4) There is a wide variation in the operational holding size of sample households. The highest number of holding area lies between 2.5 and 5 acres, and it is also found that about 41% of the farm holding are within this size class for the Northeastern region as a whole. In the holding size class categories, 0.5–1.5 and 1.5–2.5 acres, about 38% of sample households existed across the states in the region. Interestingly in the minimum size category of holding, i.e. 0.5–1.5 acres 301 farm households distributed across the region and this is the highest number followed by 299 farm households in 2.5–3.5 acres and 298 in 3.5–5.0 acres size of operational holdings. These findings from the sample data is in tune with the findings from the secondary data.
(5) For understanding the awareness level of the farmers in the region about climate changes and other associated factors the perception analysis were carried out by collecting the information on a set of questions from the respondents. The perceptions of villagers regarding variations in the pattern of cropping during the last ten years were analysed. Almost 62% of the sample farmers for the Northeastern region as a whole said that they had perceived changes in cropping pattern as a reaction to changes in climatic conditions. In this respect all respondents from Meghalaya followed by 94% in Mizoram, 85% in Manipur, 74% in Nagaland and 69% in Tripura believed that they perceive changes in the cropping pattern. However, 98% of farmers in Assam and 86% in Arunachal Pradesh did not perceive any change in the cropping pattern.
(6) The sample respondents in the region across different states revealed that they have heard about climate changes from different sources. It is to be noted that more than 61% of the sample households in the region have started to come to know about climate change from their neighbours followed by 59.6% from television, 47.6% from newspapers, 38.3% from health workers and 32.2% from the weekly magazine.
(7) About 67% of sample respondents indicated that they think the climate is gradually changing, while 9% did not answer for any specific opinion. Then we asked them about their opinion on specific forms on climate change like change in temperature, rainfall, wind, drought and humidity pattern. About 81% of the respondents in the region are of the opinion that there is an increase in temperature followed by 73% observed changes in rainfall, 51% answered

changes in wind followed by 14% for drought occurred, and 38% commented for changes in humidity pattern.

(8) To understand the causes of climate change as perceived by the respondents, a set of options were given to them to indicate their perceptions and understanding of climate change. Around 53% of the households identified deforestation as the main cause of climate change, while 23% perceived shifting cultivation as a reason behind climate change. About 46% perceive industrialization/urbanisation as reasons for climate change, while 47% indicated automobile emission as the cause. Over-population as the reasons for climate change was indicated by 42% of sample respondents.

(9) Regarding vulnerability as perceived by the sample respondents, 3.5% referred it as highly vulnerable followed by 29.6% as vulnerable and 34.4% as less vulnerable. However, one third of the sample respondent's i.e. about 32.5% preferred not to answer this question.

(10) Only 15% of sample respondents have discussed the various adaptation measures that could be adopted as a safeguard against climate change, and 20.3% reported they did not discuss it ever. While 29.4% respondents indicated that they discuss adaptation strategies sometimes, a major percentage of them, i.e. 35.3% preferred not to answer this question. It is found that overall in the Northeastern region, 36.4% households have adopted any crop management measure, 33.5% taken measures of water management and 33.2% of them have taken land management measures.

(11) As per the data from 82.2% across the region perceived lack of information on climate change followed by 73.8% for economic constraints and 70.8% for lack of extension services as a hindrance to the adaptation of climate change.

(12) The composite index approach was used to estimate the livelihood vulnerability to climate change of people residing in different parts of rural Northeastern region of India. This approach takes into consideration of seven major components: Socio-Demographic Profile, Livelihood Strategies, Social Networks, Health, Food, Water, and Natural Disasters and Climate Variability. Each of these major components comprises several indicators or sub-components.

It has been observed that values for each component and sub-component of livelihood vulnerability index varied noticeably across districts of NER which provided insight into the design and implementation of district-specific coping strategies for household. As per this study, Dhalai in Tripura, Phek in Nagaland, Mamit in Mizoram, Ri-Bhoi in Meghalaya, Kangpokpi in Manipur, Dima Hasao in Assam, and Kurum Kumey in Arunachal Pradesh are the most vulnerable districts in respective state. This observation provides policymakers, planners and development agencies with a scientific instrument to comprehend the role of physical, populations and social factors to climate change vulnerability in different parts of Northeast India.

10.3 Policy Suggestions

On the basis of the findings of the study, the following policy suggestions are recommended for sustaining livelihoods of the people and strengthening the agricultural sector in the region.

(1) The land-intensive quasi-primitive method of shifting cultivation using labour as the strategic factor of production has been evolved in the tribal society on the assumption of high land-man ratio that is no longer valid today. In fact the region appears to be over populated in the ecological sense given the land-intensive nature of agricultural production and preponderance of the primary sector in occupational structure. In the circumstances, there is the need for an environment friendly change in technology in the primary sector. It is regrettable to note that no effective alternative to 'jhumming' has yet been evolved in the region.

(2) The inhabitants of the hills in different states in the region are mostly engaged in shifting cultivation for their subsistence. They gain much less from this primitive occupation. However, as the 'jhumming' is associated with a range of serious problems that lead to imbalance in the ecosystem and endanger environmental sustainability, some alternatives should be adopted for the survival of the jhumias as well as reinforcing their current status. Though Jhum cultivation is rooted in the customs and beliefs of rural people in Northeastern region of India but it is possible to shift more and more people towards Terrace cultivation. This is reflected from the field visit of the Khonoma village. There is a need to educate people about the benefits of terrace cultivation which requires relatively less time both for sowing and harvesting. Secondly, terrace field is easy to protect from the menace of wild animals. Jhum cultivation as it is generally perceived need not necessarily be a threat to climate change. If a sustainable form of Jhum cultivation practice is encouraged by planting Alder trees in the fields, such agricultural practice may well be climate friendly and be helpful in maintaining a balanced forest ecosystem.

(3) The potential agricultural growth strategies in such areas should be related to high-value crops in horticulture, floriculture, etc. As these are likely to have high export potential, a greater step need to be taken to improve the economic status of both the jhumias and that of the region. Adequate intervention in these respects will enable the indigenous farming community to improve the adaptive capacity against the different sources of risk in livelihood vulnerabilities.

(4) Community initiative through institutional efforts has been found to be suitable alternative to allowing private property rights in shifting agriculture lands. The fact that tribal village councils and community in general are aware of consequences of unsustainable shifting agriculture, they are trying to find out suitable alternative for sustainable agricultural practices. Suitable technology should be provided to the people so that they can adopt alternative sustainable strategies. Here government extension agencies may play a very significant role.

(5) The evolution of a suitable institution for collective action for management of common property resources has been the framework suggested by Ostrom (1990). She predicted that the tragedy of the commons will not always repeat itself, that sometimes groups can find rules/institutions that enable them to manage common resources sustainably. It has been observed that community initiative through institutional efforts is an appropriate alternative to allow private property rights in areas where shifting agriculture is still prevelant. Tribal village councils and communities in general have realised that shifting agriculture is unsustainable, and many of them are looking for a suitable alternatives for sustainable agricultural practices. Therefore, there is a need to offer suitable scientific and modern technology to the people of tribal communities so that they can easily adopt alternative sustainable strategies. Here, government extension agencies may play a very important role. In his model of CPR, Dasgupta (2005) has shown that privatisation of the grazing land and cooperation over the use of that land as a CPR involves trust. Community ownership and distribution of jhum lands by village council is acting efficiently on the perception of mutual trust. On the basis of mutual trust and confidence, the traditional community institutions have grown over the centuries. These institutions are now sceptical about unsustainable nature of shifting agriculture and they are aware of its negative imapct on ecology and environment. The case of Khonoma village in Nagaland testified community concern for adopting sustainable practices.

(6) The National Forest Commission, Govt. of India has included in its advice to the Government stating that security of land tenure for shifting cultivators for both the agricultural and fallow phases should be increased. This can be done by reconsidering the classification of shifting cultivation areas and categorising them as agricultural land with adaptive forest management in the fallow period. The commission also asked for strengthening and capacitating customary institutions for improved local level governance, management of tribal, community-based natural resources, and tenurial access and control. Farmers should be allowed to clear the fallows when it is time for the cropping phase, and the species selection should be adjusted towards soil conservation as well. Effective public policy need to be devised for strengthening traditional community institutions for management of common property resources.

(7) Though the livelihood vulnerability values for each component and subcomponent varied noticeably across districts, yet the pattern is almost similar across the districts. Among the seven components, we observed that rural people are highly vulnerable in social network, livelihood and health. In social networking, the component called "government-public interaction" looks highly poor. Most of the household reported that they have not asked their local government for any assistance in the past 12 months. This clearly shows that the relationship between government and general public is at stake in rural areas of Northeast India. Since social networking is an important dimension of livelihood vulnerability, therefore, governments should create friendly environment for general people to share their problems and constraints. Secondly,

we found that livelihood component of vulnerability is immensely influenced by over dependence on agriculture. Since agricultural activities depend on a lot on natural factors, therefore, any change in rainfall and humidity affects earnings from agriculture. Thus, there is a need to reduce over dependence on agriculture by creating alternative employment opportunities in the rural areas. Thirdly, our analysis reveals that rural people spend too much time in traveling to nearest health facility centre mainly because of poor infrastructure. Therefore, government needs to carefully invest in rural infrastructure especially with regards to health infrastructure.

(8) The analysis of annual data of rainfall revealeds that Assam, Manipur, Mizoram, Nagaland, Tripura, and Sikkim have experienced decreasing trend of rainfall during the study period. Arunachal Pradesh is the only state where the annual trend of rainfall is on the rise. Similarly, strong warming trends were detected for all the states of Northeast states. With decreasing trend of precipitation, and significant increasing trends in temperature, particularly during summer and spring, the NER will face increased evaporation rates in near future. This may lead to increased aridity in the region and will put more pressure on surface and ground water resources in the study area. In the wake of such a shift in climate in the region, there is an urgent need for re-evaluation of the agricultural practices. Afforestation programme should be taken up by the community, Institutes, NGOs, schools and Governments on a routine basis. The agricultural practices that protect natural resources and has the advantage to delay effect of stress should be given importance and necessary policy or subsidy may be encouraged.

(9) An effective measure to support tribal communities in recovering and building on their traditional structures for forest management is necessary. Due emphasis should be given to develop appropriate knowledge- sharing and extension service mechanisms for rural communities for development of sustainable economic activities like diversified agriculture etc. The Evolution of collective action structures is a careful review of the historical regulation of common goods. There is a necessity of a method of governance from this examination that is analytical in nature (Ostrom, 1990).

(10) Effective institutional measures need to be implemented for capacity building and skill orientation of the jhummias in the region. Emphasis should be given to orient the younger generation in the transformation process.

References

Dasgupta, P. (2005). Common property resources: Economic analytics. In *1st Teaching Workshop on Environmental Economics for the Middle East and North Africa*. ICTP, Trieste, Italy, December 5–16, 2005.

Ostrom, E. (1990). *Governing the commons: The evolution of institutions for collective action*. Cambridge University Press.

www.ingramcontent.com/pod-product-compliance
Ingram Content Group UK Ltd.
Pitfield, Milton Keynes, MK11 3LW, UK
UKHW022210051026
15624UKWH00001B/102

* 9 7 8 3 0 3 1 5 4 9 2 6 7 *